COST ENGINEERING ANALYSIS

COST ENGINEERING ANALYSIS

A Guide to Economic Evaluation
of Engineering Projects

Second Edition

WILLIAM R. PARK, P.E.
Consulting Engineer & Economist
Overland Park, Kansas

DALE E. JACKSON, Ph.D.
Professor of Engineering Management
and Director of Engineering Management
Graduate Program
The University of Kansas
Lawrence, Kansas

A Wiley-Interscience Publication
JOHN WILEY & SONS
New York · Chichester · Brisbane · Toronto · Singapore

Library of Congress Cataloging in Publication Data:

Park, William R.
  Cost engineering analysis.

  "A Wiley-Interscience publication."
  Bibliography: p.
  Includes index.
  1. Engineering economy. I. Jackson, Dale E.
II. Title.

TA177.4.P37   1984   658.1′552   83-23474
ISBN 0-471-87671-2

Printed in the United States of America

10 9 8 7 6 5 4 3 2 1

Preface

Marketing generates the revenues;
engineering produces the profits.

A basic fact of business—though perhaps not generally recognized—is that a company's sales/marketing function is broadly responsible for bringing in the firm's revenues, and that it is up to the engineering department to design and/or produce the item or service that can be sold at a profit. The link between marketing people and engineers is through the product; marketing is consumer and product oriented, whereas engineering is product and cost oriented.

Essentially, marketing defines the consumers and their needs, with the sales force carrying dollars in the front door, while engineering spends the money and provides the services or designs the products that production pushes out the back door.

Marketing efforts show up on the firm's operating statement as a cash inflow; engineering appears as a cash outflow. It is the engineer who must exercise control and restraint on the firm's cost and expense dollars to maintain an economically acceptable spread between the cash in and cash out. As Thomas Carlyle expressed the situation, "There are but two ways of paying a debt: increase of industry in raising income, or increase of thrift in laying out."

Since the business's operating profit and cash flow position are direct functions of cash in and cash out, the economic importance of the cost engineering function is not difficult to visualize.

The main purpose of this second edition of *Cost Engineering Analysis* (as was the primary goal of the first edition, published more than a decade ago) is to ensure that technically feasible engineering projects will also be eco-

nomically attractive business ventures. This new edition is a thoroughly re-
vised and updated version. Although the basic techniques used in the economic evaluation of engineering projects have not changed since
publication of the first edition in 1973, there have been some significant
changes in the rules by which the game is played, along with improvements
in the mechanics of problem-solving.

Certainly, the most important rule changes were brought about by the
Economic Recovery Tax Act of 1981 (ERTA). At the same time, the changes
have been made far easier to accommodate by the development and proliferation of the microcomputer and its problem-solving applications soft-
ware, particularly the electronic spreadsheet.

More than ever before, good engineering requires much more than just
technological expertise. The engineer must contribute to management decisions regarding the allocation of company funds to achieve the greatest
possible benefits, whether in terms of profitable projects, salable products,
or marketable services. Engineering without economics is a poor excuse for
engineering and does not contribute to the health of a business. Maintaining
the financial strength of a company is as important a responsibility of the
engineer as maintaining its technological strength. An understanding of
markets and a knowledge of project economics, along with the ability to
understand and work closely with sales and marketing specialists, accounting and financial personnel, customers, and others involved in the success of
the project, are important qualifications for a competent engineer.

The authors hope that this new edition of *Cost Engineering Analysis* will
accomplish at least three important objectives:

1.   Provide practicing engineers with the necessary economic tools for
     analyzing projects and for preparing sound, easily understood in-
     vestment proposals and business plans.

2.   Provide management with the economic background necessary to
     thoroughly understand and properly evaluate investment proposals
     for engineering projects, so as to establish a more objective and real-
     istic basis for decision-making.

3.   Promote a better understanding among the various disciplines essen-
     tial to the successful completion of engineering projects: engineers,
     accountants, sales and marketing specialists, and other managers.

This volume presents the basic tools of economic analysis necessary for
evaluating engineering-based proposals, regardless of the industry, product,

or service being considered. It shows how sound economic principles, applied to the analysis of engineering projects, can identify the most attractive alternative available under the specified conditions. With equal importance, it shows how neglect of these basic principles can lead to costly, uneconomic, misleading, and sometimes financially disastrous results. Throughout, *Cost Engineering Analysis* stresses the practical applications of economic analysis from the engineer's viewpoint, using more than a hundred tables and figures to clearly illustrate these applications.

The first two chapters present the necessary background information regarding the functions and objectives of engineering economic analysis, the steps involved in making feasibility studies, and the essentials of preparing and presenting feasibility reports. The time value of money is explored in considerable depth (including the use of interest formulas and tables) and economic and other outside factors that influence interest rates are discussed.

Chapters 3 and 4 deal with the different approaches to analyzing and evaluating prospective investments, covering the reasoning behind investment decisions, various criteria for evaluating investment alternatives, and different ways of computing return on investment. Considerable emphasis is placed on the analysis of cash flows.

The major fixed costs related directly to capital investments—depreciation (as reflected to annual recovery deductions) and the cost of the capital employed in the project—are discussed in Chapters 5 and 6. Methods of estimating economic life and recovering capital expenditures are covered, as are approaches to estimating the costs of equity and debt capital. Major provisions of the Economic Recovery Tax Act of 1981 relating to the economic evaluation of engineering projects are discussed in detail.

Estimating, analyzing, and allocating costs and expenses make up the subject of Chapters 7–9. Incremental costs, operating and capital cost budgeting, cost/benefit analysis, short-cut estimating techniques, and a variety of other cost-related topics are covered.

Different approaches to identifying and dealing with the risks and uncertainties of the future, always present in engineering projects, are presented in Chapters 10 and 11. Some of the special tools and techniques of the professional engineer-economist useful in business and economic forecasting (risk evaluation, profitability modeling, and business planning) are disclosed in these chapters.

Chapters 12–14, unlike the preceding parts of the book, deal primarily with after-the-fact analysis, which is useful in improving the results of already-completed projects. Primary emphasis is on the analysis of financial statements—both the balance sheet and profit-and-loss statement—and on applications of these analyses in breakeven and profit planning.

The final chapter shows how broad-scope economic models can be developed to bring together all of the factors discussed previously. These models fit together the markets, prices, costs, and capital requirements into cash flow and profitability models that effectively and accurately depict that overall economic environment and competitive conditions under which a company can expect to operate in an otherwise uncertain future.

WILLIAM R. PARK, P.E.
DALE E. JACKSON, PH.D.

Shawnee Mission, Kansas
March 1984

Contents

1

Economic Feasibility: A Method, A Study, and A Report

I get the facts,
I study them patiently,
I apply imagination.

Bernard Baruch

Engineers are frequently called upon either to recommend a given project involving substantial expenditures or to submit even-handed analyses of several projects for selection by upper management. With the trend toward automation—resulting in larger initial investments and correspondingly large fixed or capital costs—the economic wisdom of investment decisions is increasingly important. Economic analysis plays an essential role in virtually every engineering project and its associated management decisions.

1

THE ELEMENTS OF ENGINEERING-ECONOMIC ANALYSIS

Good management consists primarily of making wise decisions; wise decisions in turn involve making a choice among alternatives. Engineering feasibility studies determine whether a project can succeed technically and identify the alternative ways in which that success could be realized. Economic considerations largely determine the project's desirability, whether it should be done, and, if so, where, when, and how. Feasibility studies, therefore, determine either the *which* or the *whether* of a proposed project: *which* way to do it, or *whether* to do it at all.

The term "feasible" means "capable of being dealt with successfully." In an engineering sense, feasibility means that the project being considered is technically possible. Within that possibility, engineers recognize that elements esential to the technical success of a project may range from the quickly and commercially available to being state of the art. Hence, the engineering manager must decide when too many critical elements of a project are state of the art or "pushing the frontiers of science."

Economic feasibility assumes the technical possibility that a project can be carried out and further implies that the project can be justified on an economic basis. Economic feasibility recognizes and tries to quantify the risks associated with the degree of commercial availability of the technical elements required for successful conclusion of a project. Economic feasibility measures the overall desirability of the project in financial terms and indicates the financial superiority of one approach over others that may be equally feasible in a technical sense.

ECONOMIC FEASIBILITY STUDIES GUIDE USES OF FUNDS

The main objectives of a conventional economic feasibility study are to identify and to evaluate the economic outcome of a proposed project so that whatever funds are available can be used to the best (or at least to good) advantage. The report of an engineer's economic feasibility study is frequently used as the basis for obtaining funds for financing public works projects or for allocating or borrowing funds by industrial firms. Used in these ways, the engineer's feasibility report must provide, in a quickly and readily understandable form, all data needed for top management to reach a sound decision regarding the disposition of large sums of money.

Regardless of the type of project being considered, the feasibility study

invariably requires that estimates be made of cash expenditures and economic benefits over some period of time. That period starts with the first
expenditure for depreciable, capital assets and extends for several years into
the future. For public works programs, the time period may correspond to
the useful service life of an installation. By contrast, a private company may
specify a much shorter study period.

Whatever its application, the feasibility study is always made from the
viewpoint of whoever is spending the money and usually involves the monetary comparison of alternatives. Even when only one approach to a problem is identified, an alternative still exists—that of not doing the project at
all.

FEASIBILITY STUDIES RELY ON THE SCIENTIFIC METHOD

Most engineers can recall the "scientific method." It involves observation,
problem definition, formulation of a hypothesis, experimentation, and veri-
fication. A similar sequence of seven clearly defined steps is involved in
making a thorough economic analysis:

1. Understand the problem (or recognize the opportunity.)
2. Define the objectives.
3. Collect the data.
4. Interpret the data.
5. Devise alternative solutions.
6. Evaluate the alternatives.
7. Identify the most attractive alternatives.

Two other steps are related and sequential: implementing the best alternative and monitoring the result. Those two steps, properly handled, constitute a project-based business plan.

Steps 2–7 in the given list above make up the conventional feasibility
study which consulting engineers perform for their outside clients. The client company management is usually aware that a problem or an opportunity exists before it requests a proposal for a feasibility study. On
completion of the study, the company usually implements the preferred alternative and monitors the financial results. In such cases, the control ex-
erted by the client over the implementation step may be such that the

engineer who prepared the study is able to disavow all accountability for the success of the project because of decisions made by the client management. When, however, an engineer prepares a study for internal use, she or he is much more likely to be held accountable for at least the "broad brush" results of the study.

Although the same general sequence of steps is involved in nearly all feasibility studies, many of the first seven steps overlap and are carried on concurrently. Figure 1.1 shows a general time sequence chart for typical feasibility studies.

Understand the problem

Much time and money have been wasted solving problems that did not need to be solved or devising means for taking advantage of an opportunity that was either nonexistent or transitory. Making a mistake in the first step in any problem-solving or decision-making procedure is bad enough, but when the first step is identification, a mistake is disastrous. Nevertheless, it is in defining the task that many feasibility studies fall short. There is often a substantial difference between what a company *thinks* its situation is and its

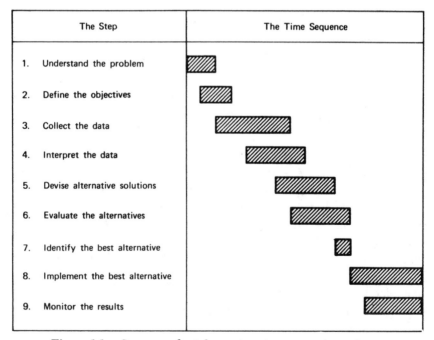

Figure 1.1. *Sequence chart for engineering economic analysis.*

real problem or opportunity. Although most problems can be solved, avoided, or lived with, several observations bear on problem solving in general:

1. Do not attempt to solve a problem without first making sure there is one.
2. If there is a problem, avoid putting a lot of effort into an attempt to solve it unless there is some reason to believe that it can be solved.
3. If there is a problem that can be solved, be certain that there is some benefit to be gained by its solution before spending too much time working on it.
4. Solve important problems or take advantage of outstanding opportunities first. There are enough big decisions to worry about without wasting time on little ones.

Define the objectives

Failure to clarify the objectives of a feasibility study has probably resulted in more unsuccessful projects and dissatisfied clients than any other single factor. Some of a firm's objectives can be defined at the outset, even before the problems are completely analyzed. A company might consider many objectives in beginning a new project. Such a plurality is acceptable so long as the objectives are clearly stated and are compatible. Some typical objectives are:

1. Achieve a specified production or sales volume.
2. Attain a specified gross profit.
3. Earn a specified percentage return on sales.
4. Earn a specified percentage return on total assets.
5. Earn a specified percentage return on investment.
6. Improve the company's performance in a specified manner.
7. Obtain a specified share of a market.
8. Solve an operating problem.
9. Pay out within a specified time period.
10. Maintain a product line.

A further classification of corporate objectives for a project is that of musts, wants, and nice-to-haves. The engineering economist is obliged to assist the client in specifying in quantitative, measurable terms the objectives of the study as well as the procedures for measurement. If the feasibility study is to determine whether or not to become involved in a proposed project, the specified objective will provide the cutoff point. If the proposed plan can satisfactorily accomplish the objective, the plan will be approved. Or, if the study is to identify which of several ways of accomplishing something is the best, the relative desirability of alternative plans must be measured in terms of how well they meet the objective criteria.

Collect the data

Data collection can begin as soon as the problems and objectives are well enough defined to indicate the types of data useful for carrying out the study. All data pertaining to the company's specific problems and objectives should then be collected and maintained in current order. Much of the data will be of a historical nature, and a good backlog of historical economic data can prove invaluable to the engineer involved in these types of studies.

The data collection phase should normally begin with a review of published literature on whatever subjects are being investigated; too much time has been wasted developing data that were already available to those who knew where to look. Whatever data are not readily available from the literature can frequently be obtained through private company sources; otherwise, substitutes for data must be improvised through assumptions and estimates. As data are collected, they should be carefully referenced so that original documents can be retrieved if needed. Also, clear distinctions should be made between factual data, estimates, and assumptions. The reliability of the conclusions drawn from data is no better than the reliability of the data themselves.

In the collection of data, economic studies—unlike technical studies—can seldom employ experimentation to develop new data. Economic factors are generally less controllable than physical conditions, and so many economic variables may be involved that their interrelationships are unclear.

Interpret the data

When most of the apparently significant data have been collected, the interpretation phase can begin. If the raw data are organized such that each bit of information, complete with references, can be found when needed and with little delay, the analysis of that data will be more meaningful.

Most data are either quantitative or can be adapted to statistical analysis. The analysis task, however, can become a burden because of the sheer bulk of the data. In such cases not only is the analysis time-consuming, but important information and relationships may remain hidden. The availability of microcomputers and data base or filing programs can lessen the likelihood of losing information, but the use of such advanced data-handling technology itself requires an investment in time and equipment. Computer programs for statistical analysis of quantitative and attribute data can sometimes pay for their consumption or resources not only by the organization of the data but by their suggestion of conclusions not obvious from the data alone.

The primary reward for data organization, classification, and analysis is to help make sense out of the mass of information available to the engineer-economist. From that treatment of data the engineer can help management make more informed decisions that are likely to be more consistent with the engineer's understanding of the situation.

To the engineer who enjoys the creative and useful manipulation of data the interpretation phase is the most fascinating part of the whole study. That fascination includes an interest not only in a quantitative description of what has happened but a mathematically sound estimate of what may happen. While perhaps not sharing the fascination with the data itself, management can profit from the interpretation phase by its provision of a sound and objective basis for translating knowledge of the past into understanding and estimates of the future.

Devise alternative solutions

The process of analyzing and interpreting data can sometimes suggest to the engineering-economist one or more courses of action. Those suggestions are the first steps at devising alternatives for solving the problems and non-certainties invariably associated with investments. In this phase of the study the primary demand upon the engineer is that of disciplined creativity. That skill may be called upon in a purely technical sense, such as in the selection of materials or proposing ways to solve design problems. Creativity can be equally valuable when applied to generating cost vs. investment alternatives in the economic realm.

Seldom, if ever, will the engineer find one economic alternative that is so clearly superior to the best that it literally demands unquestioning selection. In fact many engineers have learned by adverse experience that the identification of the clearly superior alternatives usually means one or more attactive alternatives are temporarily in hiding—put there by the engineer's own prejudices or a preconceived notion of what the client or management "really wants."

To fully meet the requirements of this phase the engineer will not ignore any possibility that might merit later consideration. Trade-offs and workable—if not always fair—compromises will inevitably follow. If the project contemplates producing and selling a product, then different manufacturing methods and various levels and patterns of sales growth merit consideration. If the project is a study of manufactured cost of a product, the possibility of changes in labor costs, labor productivity, interest rates, taxes, and other cost factors can be introduced to create economic alternatives. Any factor that the engineer sees as at all reasonable is suitable for inclusion in the evaluation.

Evaluate the alternatives

The engineer can estimate the results of alternative proposals as they are developed. If the engineer carries out this portion of the study in close coordination or at least after careful consultation with the client or management, the final proposal has an infinitely better chance of understanding and perhaps acceptance. If, for instance, the engineer learns that the meaning he/she ascribes to return on investment is different from that accepted by the client, avoiding that potential conflict will avoid failures to approve otherwise ideal proposals.

Whatever basic differences of terminology may exist, the economic desirability of a project depends primarily upon the magnitude and timing of cash flows. The basis for evaluating alternatives is sound, however, only after the engineer has learned which figure or figures or economic merit management will use to rank the alternatives. The rest of the process is the application of arithmetic and algebra to the data. Some organizations may, by habit or sensitivity to the economic winds of the time, set a minimum payout time, with or without regard to discounting the cash flow. Otherwise, the engineer can rank the alternatives in their apparent order of economic desirability.

Particularly those engineers whose recommendations affect the spending of public funds are well advised to document not only the alternative that was selected but those that were not, including the reasons for their rejection.

Select the best alternative

Selection of the best alternative must wait until all have been evaluated. Sometimes the choice is the responsibility of the persons conducting the feasibility study; more likely, the choice is a top-management decision in which selection is made from among the best few choices. Most of the alternatives originally developed in the fifth step will probably not be particularly desirable after objective evaluation, and the choice will be narrowed down to just a few "best" alternatives. From this point, the decision regarding which course of action to follow may be extremely difficult. Presumably, the only remaining choices are among the relatively good ones, so the worst that can happen is that a good—although not necessarily the best—choice is made. An *acceptable* choice, though, is almost a certainty if the analysis has proceeded as outlined here.

Selecting the best alternative requires identifying the one that comes closest to accomplishing the firm's desired objectives. There may be cases in which *none* of the alternatives meet the objectives; this indicates either a poor choice of alternatives or a poor choice of objectives, and one or the other must be changed. But one plan is bound to come *closest* to accomplishing the objectives, and if action is required, this plan offers the best *available* choice.

The main purpose of a feasibility study may be to recommend a particular course of action or to present without prejudice the results of analyzing several courses of action. In either event, the feasibility study should objectively and quantitatively show the economic results and implications of following differect actions. Whether further action is taken will usually depend upon factors and policies beyond the engineer's knowledge, control, or accountability.

Plan the implementation of the selection

As soon as evaluation is complete and the best alternative (or alternatives) has been identified, the feasibility study goes into administrative channels. The engineer's role in this phase may ultimately range from complete responsibility for design and construction to almost complete inactivity. In either case, those who prepared the study and presented the plans for management consideration should participate in the implementation phase at least enough to insure that the proposed plans are the ones being implemented. Also, any serious deviations from the original estimates should be quickly noted, and their impact on the overall attractiveness of the project should be appraised and reported.

Monitor the results

The moment of truth arrives when the engineer's estimates are compared with the actual results of having acted on his analysis and findings. The effectiveness of any proposal can be measured only in terms of how well it accomplishes the objectives. If the objectives and goals have been adequately defined, the measurement is a relatively simple task. Some important benefits can accrue to the engineer who carefully monitors the results of projects and compares them with the results in his estimates. Perhaps the greatest benefit is an improved estimating technique for future studies; another is the added care an estimate will be given, knowing that it will be viewed later in the light of reality. Inasmuch as a company can choose among project alternatives by using several economic evaluation techniques, the engineering economist who serves that company can improve the quality of that service by monitoring more than the economic results. The engineer can record the actual attainment of economic estimates that were subject to risk analysis. The engineer can also record, given sufficient data, the actual internal rate of return on a project, even though another method of comparing projects was chosen by the client.

THE FEASIBILITY REPORT

The bridge between the feasibility study and the implementation of one of the alternatives that it develops is the feasibility report. Such a report may be intended either to inform its readers of the results of an objective analysis of several alternatives or to persuade its readers to approve a given course of action. Regardless of how thoroughly a project engineer has investigated a project, the report will reflect that thoroughness and conviction effectively only if it is couched in language familiar to the person responsible for the final decision. It should also summarize the essential information in a minimum of space, preferably within one, double-spaced page.

One of the most difficult assumptions for an engineer to overcome in report writing is that the readers are about as well-informed and at least equally interested in the scope of activities that went into the study. On the contrary, most readers are not as well-informed as engineers, who have both technical and economic skills. Furthermore, none of those readers will set aside or even have the time to read and comprehend all of the material within a report.

In these days of word processors and computers that know how to spell, there is little excuse for errors of spelling, punctuation, and capitalization. Decision makers whose early education or personal interests have developed in them a sense of correct English grammar and usage are likely to suspect the quality of the engineer's study when the report of it contains language errors.

The authors are aware of at least 3 dozen books to help engineers write better reports. Any library or bookstore contains at least a few of these, and the reader is commended to such sources. Most of the books on technical writing stress errors in style, usage, and grammar. They also give tests for readability that involve counting the average number of words in sentences and syllables in words. The authors' experience has led them to conclude that by the time a report has been redrafted, the editing of such a draft is at least as difficult as writing it in the first place. The old saying of quality control managers that, "You can't inspect quality into a product," is applicable to reports. The engineer who writes the report and the supervisor who tries to edit it will both save time if the engineer starts with short sentences that express one or at most two thoughts and keeps the numbers of syllables in words as low as possible, preferably averaging less than three. Books on writing for engineers also recommend report formats. Those formats can be divided into front matter, body, and back matter.

Front matter

The most important part of the front matter and, indeed, of the entire report is the executive summary, which is a report-in-miniature. It tells what is proposed and why it is worthwhile in economic terms. Managers whose assignments call upon them to make decisions about capital investments are busy people and may read no further than a cover letter (see below) and the executive summary. Such managers are also known to request the conclusions or recommendations at the start of an oral presentation and to leave the meeting as soon as they have heard those elements. The front matter also includes the report cover, a title page, a preface, acknowledgments, a table of contents, a list of illustrations, a list of tables, and a list of appended matter.

Body

Here the engineer is less obliged to conserve space and to be meticulously accurate in the use of language, but the need for brevity continues. The emphasis in the body of the report should be upon logic, clarity, and accuracy. Those goals are most easily reached if the writer starts with a good outline. Engineers who start with an outline of short, complete sentences will find that it helps them organize their thoughts and provides ideal first sentences for each paragraph within the body of the report. A uniform system of numbering, indentation, headings, and presenting supporting graphs and tables will make the report of value to those who later implement its findings.

Back matter

Appendices, graphs, tables, perhaps a subject index, and the back cover form the rest of the report. An appendix is intended to include data and other detailed information of possible technical interest, but is not necessary for a manager to accept or reject the findings of the study. The engineer may be tempted to use a graph whenever confronted with dependent and independent variables and to assemble data in one or more tables. Such practices are not always helpful to a reader but can be made so by observing this rule: always give a graph or a table a title that summarizes the message it presents. It is redundant to entitle a graph "The Effect of Transmission Rate on Volume," since this information is contained in the labels of the ordinates. Similarly, uninformative titles for tables require the reader to spend time comprehending the labels on the columns and rows, whereas a title such as "Transmission Data Show Thickness Effect to Peak at 3 Inches" lets the reader decide quickly whether or not to examine the data in detail.

LETTER OF TRANSMITTAL

Such a letter may accompany the feasibility report separately or may be bound into the front of the report, just after the cover. The transmittal letter informs the recipient that the report is either in the envelope or is being sent under separate cover. The letter identifies the subject of the report, cites the contract or agreement under which the study was conducted, and refers the recipient to those individuals who are accountable for the preparation of the report.

GENERAL FORMAT

Figures 1.2 and 1.3 illustrate the format and contents of typical economic feasibility studies. The reports from which these Contents were taken contained about 40 pages.

The study described in Figure 1.2 was an analysis of a proposed total energy system for an apartment complex. The reader of this report would have had an easier time anticipating its content had the headings of the subsections been short, headline-like sentences. Section 6 would have been strengthened by a subsection on taxes. Section 7 needs a subsection on inflation. The conclusions and recommendations in Section 9 belong in the executive summary. Had the study been intended to lead to a recommendation, it would have made its point more effectively if its conclusions were

Summary

I.   Introduction
II.   Comparison of alternative energy systems
III.   Determinants of economic feasibility
IV.   The apartment project
    A.   General description
    B.   The proposed energy systems
V.   Analysis of energy system operations
    A.   Electric loads
    B.   Equipment utilization and fuel consumption
    C.   Heating–cooling system fuel requirements
VI.   Estimated costs and revenues
    A.   Capital costs
    B.   Operating costs
    C.   Operating revenues
VII.   Economic feasibility
    A.   Present worth
    B.   Return on investment
    C.   Payout time
VIII.   Effect of economic variables
    A.   Timing of future construction
    B.   Variations in occupancy rates
    C.   Variations in net annual savings
IX.   Conclusions and recommendations

Appendix

Figure 1.2. *Feasibility of a Total Energy Installation.*

Introduction

    A.   Purpose
    B.   Scope of study
    C.   Historical background

Summary of findings and recommendations

    I.   Demand for water
        A.   Average day demand
        B.   Maximum day demand
        C.   Maximum hour demand
        D.   Summary of demand rates
    II.   Existing water system
        A.   Supply
        B.   Storage
        C.   Distribution
    III.   Proposed Improvements
        A.   Supply
        B.   Storage
        C.   Distribution
    IV.   Estimated Costs
        A.   Short–range program
        B.   Long–range program

Appendix

Figure 1.3. *Report on Water System Improvements.*

buried within the subsections of the body. If, on the other hand, the study was intended to present a series of alternatives, conclusions are sufficient; recommendations have no value.

Figure 1.3 describes a study of proposed improvements for a municipal water system. In this case, some action is required. The purpose of the study is to determine the most economical alternative for satisfying the physical demands placed upon the system.

FOR AN ORAL PRESENTATION, START WITH THE ANSWER

The engineer who assumes that she or he is not speaking to a captive audience will better appreciate the value of presenting the most significant conclusions or recommendations as early as possible.

One of the most popular sales presentation techniques—dating back at least to the 1920s—is called the AIDA approach, with the initials standing for Attention, Interest, Desire, and Action. By following this approach, an effective presentation to an audience (either a group of people or an individual) can accomplish these four objectives:

1. Attract *attention* to the presentation.
2. Arouse *interest* in the subject.
3. Generate a *desire* for the project, product, or service.
4. Motivate the desired *action.*

The last three steps in the AIDA presentation can be related to the previously mentioned three-step approach to technical reporting. *Interest* can be aroused by briefly describing and emphasizing the most important (and therefore the most salable) points to be covered in the presentation. *Desire* for the project should evolve naturally during the listening to or reading of the presentation itself, as facts are first presented, then interpreted, finally evolving into meaningful and relevant conclusions. *Action* results from translating the conclusions into positive recommendations for what needs to be done.

It is the *attention* phase that is often left out of the engineering proposal—where the person or organization authoring the report or making the presentation must establish a clear relationship between himself, the subject of the presentation, and the audience to which it is directed. Specifically, he owes his audience some justification for:

being here, either in person or represented by a written document,

presenting this subject,

to this audience,

at this time.

The audience's attention is earned only if these points are adequately clarified. The first 30 seconds or 50 words of either a written or oral presentation must command the initial favorable attention that will make the reader or listener *want* the presenter to continue his story.

The first 50 words in a presentation are in fact more important than the next 10,000 words in terms of their overall impact on the audience. There is no second chance to create a first impression.

A good "human engineer" knows that if he is unable to establish a firm relationship between himself, the subject, and the audience within the first 30 seconds of his presentation the audience may be lost forever—mentally, if not physically. It is these 30 seconds or 50 words that either provide a good springboard from which to jump into the formal presentation, or impose a handicap upon the rest of the presentation from which it may never recover.

Some final words of advice concern the means of losing the effects of an otherwise satisfactory presentation. The engineer, in those cases where she or he has made a recommendation will usually encounter objections. Those objections may be logical, emotional, or some of both. If they are logical and are delivered with little or no nonverbal transmissions that may suggest underlying emotion, they can usually be answered with data or analyses of data. If the objections are logical but have already been answered in the presentation, or if the objections are emotional, the best answer is no answer at all. On the contrary, the best answer is taken from the practice of client-centered therapy, as developed by Carl R. Rogers.

For example, suppose an engineer's presentation is met with an objection having to do with the risk associated with sales volume in the third year of operation. Suppose also that the question followed a presentation in which that datum was shown to have little or no effect on the recommendation. The engineer may be tempted to point out that the objection was dealt with in the presentation or even to repeat that part of the material. The engineer, however, is far more likely to achieve understanding and agreement if she or he responds with, "Sounds as if you are concerned about how far off that estimate could be." The more emotionally based an objection appears to be, the more valuable it is to the engineer to use that "reflective" technique until the objectors have had their say. If the engineer follows that discipline carefully, someone in the audience will inevitably find something good about the proposal. At that point, the engineer can ask the speaker to expand on whatever it is that sounds good. The almost certain result is acceptance, assuming that the study was done properly in the first place.

SUMMARY

The growing capital costs associated with engineering projects have increased the necessity for sound engineering-economic analysis. An economic feasibility study determines either *which* of several ways a project should be carried out, or *whether* the project should be carried out at all.

The main objective of the feasibility study is to predict the outcome of a proposed expenditure in financial terms so that available funds can be put to their most advantageous use. Seven steps comprise a complete engineering economic analysis: (1) understand the problem, (2) define the objectives, (3) collect the data, (4) interpret the data, (5) devise alternative solutions, (6) evaluate the alternatives, and (7) identify the best alternatives. A follow-up business plan then calls for implementation of the best alternative and monitoring of the results. The first seven steps make up the usual feasibility study and form the basis for a report to management. The first page of that report, following the letter of transmittal, is most useful to top management if it summarizes the important features of the proposed project. It either helps to generate the desired decision or provides for an unbiased consideration of several alternatives on their merits.

2

Borrowed Money, Interest Rates, and Time

The human species, according to the best theory I can form of it, is composed of two distinct races, the men who borrow and the men who lend.

Charles Lamb

Since most business ventures involve the use of other people's money, and usually for several years, interest—the rental charged for the use of borrowed money—and the time over which it is paid play an important role in the engineer's preinvestment analysis. From a borrower's standpoint, the opportunity to invest borrowed money at a higher rate than must be paid for its use justifies the payment of interest. From the lender's standpoint, interest is compensation for obtaining the funds from other sources and making them available to the borrower. These, then, are the two primary reasons for having interest: the borrower's opportunity to invest money and the lender's desire to profit from the loan.

BORROWED MONEY COMES FROM TWO FINANCIAL MARKETS

The financial markets from which money is borrowed can be broadly classified into two categories: (1) the capital, or investment, market and (2) the money market. However, capital and money markets often overlap and the time for which funds are committed is probably the best way to distinguish between them. The money market usually refers to funds borrowed or loaned for a year or less, whereas the capital market encompasses longer-term obligations. Overall, savings institutions provide about half the total funds available in the financial markets, commercial banks contribute about 30 percent, and the remainder is supplied by business corporations and individuals.

LONG TERM LOANS COME FROM THE CAPITAL MARKET

The capital market, from which most investment-type funds are obtained, includes:

1. Corporate bonds.
2. Corporate equities.
3. Mortgages.
4. Commercial bank term loans.
5. Small Business Investment Companies.

In the capital market, mortgages account for more than half of all funds used. More than two-thirds of the funds employed in the capital market are obtained from savings institutions.

Bonds are long-term obligations, have a specific value, and offer assured interest at an established rate. Bonds, to be attractive to investors, must be heavily secured by assets, anticipated revenues, or other means. Bond repayment schedules are usually set up to pay interest periodically—one to four times a year—with the principal amount payable upon the bond's maturity. Most bonds are sold on the primary market (when newly issued), although there is also a large secondary, or resale, market for many of them. Zero interest bonds have recently become popular. The bondholder's profit is realized at maturity.

This situation is just the opposite of the equity market, which has a small primary sale on new issues but a huge secondary market (the "stock market"). Thus, the equity market makes up a relatively small portion of the total funds available for corporate financing. Corporate equities, or capital stocks, may be issued by firms whose cost of capital would be lower than could be obtained by entering the bond market, or whose financial position is such that there would be insufficient market for their bonds.

Mortgages are long-term securities given for repayment of a debt usually connected with real property. The mortgage is secured by title or lien on the property being financed. Repayment of a mortgage is usually made in a series of uniform amounts, with each payment including both principal and interest.

Commercial bank term loans may be used for start-up capital or for expansion. The borrower agrees to a level payment schedule, usually monthly, with the early payments consisting mostly of interest. In some instances, the U.S. Small Business Administration (SBA) may be willing to guarantee up to 90 percent of the loan, which means that the bank has much less of its funds at risk. In a few instances, the SBA may make direct loans.

Small Business Investment Companies are licensed by the SBA to buy a limited amount of stock in and lend to a given small business with the possibility of loan guarantees by the SBA. SBICs typically concentrate their participation within certain business areas and within a range of sizes of lenders, as measured by their annual sales. Minority Enterprise Small Business Investment Companies are licensed by the SBA to assist in the creation and growth of minority-owned business.

SHORTER TERM LOANS COME FROM THE MONEY MARKET

The money market—the shopping center for short-term (1 year or less) funds—includes:

1. Commercial bank loans.
2. Commercial paper.
3. Bankers' acceptances.

Just over half of the short-term funds are supplied by commercial banks, with about 30 percent coming from business corporations and the remainder from other investor groups and individuals. The most important single source of short-term funds is the commercial bank loan.

Commercial bank loans are made to established corporate customers at a negotiated rate of interest, "prime" or above. (The prime rate refers to the interest rate that commercial banks charge their preferred customers for short-term funds). They may be secured or unsecured, depending on the customer's credit standing and financial situation, and are issued for varying lengths of time.

Commercial paper refers to secured or unsecured promissory notes issued by corporations and sold to investors, most of whom are other corporations. These notes are written, unconditional promises to pay—on demand or after a short time—a specified amount of money. The notes are often discounted at commercial banks.

Bankers' acceptances are bills used to finance the import, export, or transfer of goods. The bank guarantees their payment at maturity. Factoring firms typically loan up to 80 percent of the amount on an invoice to a credit worthy customer.

INTEREST RATES

The money supply and inflation

Each type of loan—mortgage, corporate bond, personal loan, or whatever—carries a different price, or interest rate. This interest rate, regardless of the type of loan involved, is a function of the supply of, and the demand for, money.

When funds are in short supply relative to demand, short-term interest rates can be expected to rise. When short-term rates go up, long-term rates cannot help but be affected. Although short-term rates are determined by current supply-and-demand factors, long-term rates must anticipate supply-and-demand relationships over the future life of the interest-bearing security.

The cost, current and anticipated, of money to commercial banks naturally has a strong impact on the price they charge borrowers for the use of money. When the upper limit on the interest that banks were allowed to pay on time deposits (savings accounts and certificates of deposit) was removed, large depositors shifted their reserve funds to banks from treasury bills and other short-term securities. This transfer gave commercial banks more money to lend.

Banks make money only by lending or investing other people's money. They must, therefore, borrow from other sources if they are to have money available for loans, even if the other sources are relatively expensive. These outside funds may come either from U.S. banks having excess reserves or from European banks holding U.S. dollars. Regardless of the source, the borrowing bank will probably have to pay a substantial premium for outside funds.

Nevertheless, during periods of rapid inflation borrowers are willing to pay almost any amount to obtain money because: (1) they expect costs of equipment and buildings to go up, (2) they anticipate raising prices to cover the increased price of borrowing money, and (3) they expect to repay the loan with cheaper money. At the same time, lenders must ask high-interest rates to protect themselves from a loss in purchasing power.

The Federal Reserve Bank

Government policy is an important determinant of the rate of interest, and the direction of the nation's economy can be controlled to some extent by how the government manipulates the money market.

The Federal Reserve has the power to: (1) change its reserve requirements to member banks, thus changing the amount of money available for loans; (2) change its pattern of open-market operations, which it frequently does; and (3) change the discount rate, thereby lowering or raising the cost of money to banks. The Federal Reserve can also set an upper limit on the interest rates paid on time deposits by banks under its jurisdiction.

Even with these controls at its disposal, the money market sometimes gets away from the government, resulting in threats of wage and price controls and other strict measures to bring the economy back in line. Federal monetary policy, while important, is but one of many factors that influence the cost of money.

HIGH INTEREST DISCOURAGES INVESTMENT

High interest rates reflect a tight money supply. The resulting difficulty in obtaining money, or the high price that must be paid for it when it is obtained, show up in the nation's economy in several ways.

First, high interest rates make it more expensive to make a new investment. This is especially important for public utilities and municipalities, whose funds must be raised from outside sources through the issuance of bonds. Such fixed-income, long-term securities are generally unattractive during periods of rising prices, as evidenced by the difficulties experienced by state and local governments in marketing bonds to finance schools, roads, sewers, and other public improvements.

A second major effect of high interest rates is that, even with internally generated funds, new capital projects appear less desirable relative to the returns offered by alternative short-term investments such as U.S. Treasury bills. When capital investments must be financed with borrowed money, high interest rates reduce the effective rate of return on invested capital. If the expected profitability of a proposed investment does not compare favorably with the prevailing interest rate—a convenient standard for comparison—considering such factors as risk and liquidity, a firm will not make an investment.

In the private construction field, where mortgage financing is commonly employed, interest rates also have a strong effect. Doubling the rate of interest on a mortgage more than doubles the amount of interest paid. Raising the interest rate on a 20-year mortgage from 5 to 10 percent, for example, actually increases the total amount of interest paid by about 2.24 times; a 20-percent increase in the interest rate increases interest payments by about 23 percent. During 1982, interest rates in the high teens for home mortgages were partially responsible for the fewest housing starts in recent U.S. history.

Thus, slowdowns in home building, delays in public projects, and decreased capital outlays by large corporations all may result, directly or indirectly, from high interest rates. A general feeling of uncertainty about the immediate future of the money market also contributes heavily to an overall slowdown in the economy.

INTEREST FORMULAS

The effect of time on money

With three basic interest formulas and their reciprocals, the engineer can convert an amount of money spent or received on a given day to its equivalent amount days, months, or years earlier or later. In all these formulas, i represents the rate of interest, usually expressed on an annual basis; n indicates the number of interest periods; P is a present amount of money, or the equivalent present value of some future amount; F is the future amount of money, either payable or receivable in a single lump sum; and A is a uniform annual amount.

Figures 2.1, 2.2, and 2.3 show the factors pertaining to each of the six compound interest formulas for selected interest rates and over different time periods.

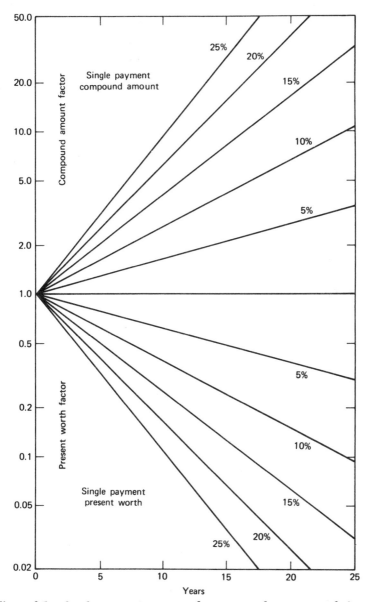

Figure 2.1. *Single payment compound amount and present worth factors.*

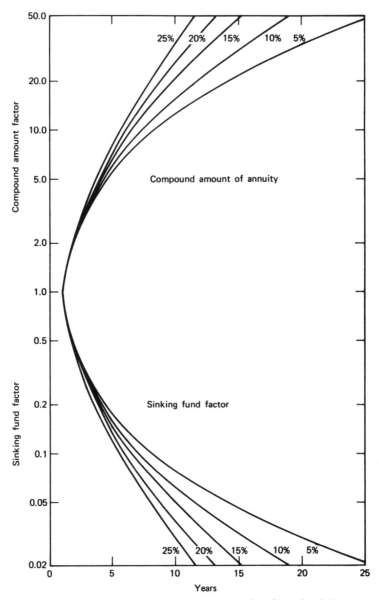

Figure 2.2. *Annuity compound amount and sinking fund factors.*

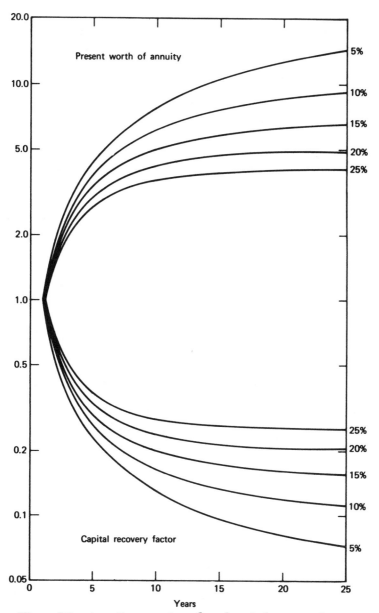

Figure 2.3. *Annuity present worth and capital recovery factors.*

27

SINGLE PAYMENT COMPOUND AMOUNT FACTOR.   Multiplying some present amount P by this factor will yield the future amount F to which P will accumulate in n years at i percent of interest.

$$F = P(1 + i)^n \tag{1}$$

The value of $(1 + i)$ is plotted in the top half of Figure 2.1 vs. time in years for a range of interest rates.

SINGLE PAYMENT PRESENT VALUE FACTOR.   This factor multiplied by a future amount F, receivable in n years, gives P the present value of F, with interest at i percent. This is the reciprocal of Equation (1).

$$P = F \left[\frac{1}{(1 + i)n} \right] = F(1 + i)^{-n} \tag{2}$$

This factor is shown graphically in the bottom part of Figure 2.1 vs. time in years for several interest rates.

ANNUITY COMPOUND AMOUNT FACTOR.   This factor, multiplied by an annual payment A, gives the future value F to which A will accumulate in n years at i-percent interest.

$$F = A \left[\frac{(1 + i)^n - 1}{i} \right] \tag{3}$$

Values of the factor are plotted in the top half of Figure 2.2 for several interest rates.

SINKING FUND FACTOR.   This factor, multiplied by a specified future amount F, gives the annual amount A that must be deposited over a period of n years at interest rate i to accumulate F. This is the reciprocal of Equation (3).

$$A = F \left[\frac{i}{(1 + i)^n - 1} \right] \tag{4}$$

ANNUITY PRESENT VALUE FACTOR.   This factor times a uniform annual payment A gives the present amount P that can be paid off over n years at i-percent interest.

$$P = A \left[\frac{(1 + i)n - 1}{i(1 + i)^n} \right] \tag{5}$$

These P values are shown in the top half of Figure 2.3.

CAPITAL RECOVERY FACTOR. This factor, multiplied by some present amount P, gives the uniform annual amount A required to amortize, or completely pay off, a loan over n years at i-percent interest. The capital recovery factor is equal to the sinking fund factor plus the interest rate. This factor is the reciprocal of Equation (5).

$$A = P \left[\frac{i(1 + i)^n}{(1 + i)^n - 1} \right] \tag{6}$$

The bottom part of Figure 2.3 shows the capital recovery factors.

Uses Beyond Interest

The primary uses of the interest formulas in calculating dollar values are apparent from the preceding descriptions. But these formulas also apply equally well to anything exhibiting a constant *rate* of growth (as opposed to constant arithmetic increases). Population growth is a good example of a fairly constant rate of increase, as are per capita usage of electrical energy or water and various measurements of production and consumption.

The single payment compound amount factor, for example, is the basic formula used in measuring growth rates. If the population of an area has grown from 1 million persons in 1950 to 1.7 million in 1970, these figures substituted in Equation (1) indicate an average compound rate of 2.7 percent annually over the 20-year period. The same formula can then be used to project the present population to any future date, assuming the rate of growth to remain constant. The compound amount factor plots as a straight line on semilog graph paper.

The four annuity-type formulas can be used whenever a uniform stream of quantities is involved—such as comparing a series of operating savings from a new piece of equipment, a new process, a different design, or an alternative energy source. These formulas measure, in different ways, the cumulative effect of future quantities and allow them to be expressed as a single equivalent amount.

The capital recovery factor is especially useful in engineering economy studies comparing alternatives having different useful service lives. The capital recovery factor amortizes the initial investment over its life, thereby reducing the alternatives to their equivalent annual costs. It considers both depreciation (principal) and interest, thus representing direct out-of-pocket costs of a project financed entirely through outside funds.

USING INTEREST TABLES

Before the proliferation of microcomputer and electronic spread sheets, a set of interest tables, giving numerical values for all six of the basic interest formulas for different interest rates and time periods was an important part of every engineering economist's library. These tables are included in most standard handbooks of mathematical tables, although their usefulness is sometimes restricted because of the limited range of values presented. In feasibility studies and investment analyses, interest rates of 25 percent or higher are often used.

Table 2.1 shows a page from a typical interest table. By using the 10-percent compound interest factors shown in the table, solutions to all typical interest problems at this rate can be quickly found.

In the first column, the *compound amount* factor shows the amount that a dollar accumulates to over N time periods. One dollar invested now at 10 percent accumulates to $2.59 over a 10-year period, and to $6.73 in 20 years.

The *present worth* factors in the next column are reciprocals of the compound amount factors; $1.00 receivable in 10 years is worth only $0.386 now; $2.59 receivable in 10 years has a present worth of $2.59 × 0.386, or $1.00.

The *sinking fund* factor in the third column shows the amount that must be invested at 10 percent each year to accumulate to $1.00 in n years. Thus $62.75 must be invested annually to accumulate, with interest, to $1,000 in 10 years.

In the next column, the *capital recovery* factor shows the annual payment required to cover principal and interest in equal annual amounts over an n-year period. A $1000 loan can be retired in 5 years by paying back $263.80 annually. A total of $1,319 would be repaid, of which $1000 is the principal and $319 is interest.

The *compound amount* factor in the fifth column shows the total accumulation, with interest, of an equal amount invested each period for n periods. If $1000 were invested each year at 10 percent for 20 years, the total amount that would accumulate at the end of the twentieth year would be $1000 × 57.275, or $57,275. In this case, only $20,000 represents the principal, the remaining $37,275 being interest.

The last column indicates the *present worth* of a uniform annual series of payments. This shows that to purchase a $1000, 20-year, 10-percent annuity requires a present payment of $8514. Hence, a project returning $1000 annually for 20 years has a present value of $8514.

Table 2.1 A Page From a Typical Interest Table

10.00 PERCENT COMPOUND INTEREST FACTORS

N PERIODS	SINGLE PAYMENT		UNIFORM ANNUAL SERIES				N PERIODS
	COMPOUND AMOUNT FACTOR GIVEN P TO FIND S $(1+I)^{**N}$	PRESENT WORTH FACTOR GIVEN S TO FIND P $\dfrac{1}{(1+I)^{**N}}$	SINKING FUND FACTOR GIVEN S TO FIND R $\dfrac{I}{(1+I)^{**N}-1}$	CAPITAL RECOVERY FACTOR GIVEN P TO FIND R $\dfrac{I(1+I)^{**N}}{(1+I)^{**N}-1}$	COMPOUND AMOUNT FACTOR GIVEN R TO FIND S $\dfrac{(1+I)^{**N}-1}{I}$	PRESENT WORTH FACTOR GIVEN R TO FIND P $\dfrac{(1+I)^{**N}-1}{I(1+I)^{**N}}$	
1	1.1000000	.9090909	1.0000000	1.1000000	1.0000000	.9090909	1
2	1.2100000	.8264463	.4761905	.5761905	2.1000000	1.7355372	2
3	1.3310000	.7513148	.3021148	.4021148	3.3100000	2.4868520	3
4	1.4641000	.6830135	.2154708	.3154708	4.6410000	3.1698654	4
5	1.6105100	.6209213	.1637975	.2637975	6.1051000	3.7907868	5
6	1.7715610	.5644739	.1296074	.2296074	7.7156100	4.3552607	6
7	1.9487171	.5131581	.1054055	.2054055	9.4871710	4.8684188	7
8	2.1435888	.4665074	.0874440	.1874440	11.4358881	5.3349262	8
9	2.3579477	.4240976	.0736405	.1736405	13.5794769	5.7590238	9
10	2.5937425	.3855433	.0627454	.1627454	15.9374246	6.1445671	10
11	2.8531167	.3504939	.0539631	.1539631	18.5311671	6.4950610	11
12	3.1384284	.3186308	.0467633	.1467633	21.3842838	6.8136918	12
13	3.4522712	.2896644	.0407785	.1407785	24.5227121	7.1033562	13
14	3.7974983	.2633313	.0357462	.1357462	27.9749834	7.3666875	14
15	4.1772482	.2393920	.0314738	.1314738	31.7724817	7.6060795	15

Table 2.1 *(Continued)*

16	4.5949730	.2172291	.0278166	.1278166	35.9497299	7.8237086	16
17	5.0544703	.1978447	.0246641	.1246641	40.5447028	8.0215533	17
18	5.5599173	.1798588	.0219302	.1219302	45.5991731	8.2014121	18
19	6.1159090	.1635080	.0195469	.1195469	51.1590904	8.3649201	19
20	6.7274999	.1486436	.0174596	.1174596	57.2749995	8.5135637	20
21	7.4002499	.1351306	.0156244	.1156244	64.0024994	8.6486943	21
22	8.1402749	.1228460	.0140051	.1140051	71.4027494	8.7715403	22
23	8.9543024	.1116782	.0125718	.1125718	79.5430243	8.8832184	23
24	9.8497327	.1015256	.0112998	.1112998	88.4973268	8.9847440	24
25	10.8347059	.0922960	.0101681	.1101681	98.3470594	9.0770400	25
26	11.9181765	.0839055	.0091590	.1091590	109.1817654	9.1609455	26
27	13.1099942	.0762777	.0082576	.1082576	121.0999419	9.2372232	27
28	14.4209936	.0693433	.0074510	.1074510	134.2099361	9.3065665	28
29	15.8630930	.0630394	.0067281	.1067281	148.6309297	9.3696059	29
30	17.4494023	.0573086	.0060792	.1060792	164.4940227	9.4269145	30
31	19.1943425	.0520987	.0054962	.1054962	181.9434250	9.4790132	31
32	21.1137767	.0473624	.0049717	.1049717	201.1377675	9.5263756	32
33	23.2251544	.0430568	.0044994	.1044994	222.2515442	9.5694324	33
34	25.5476699	.0391425	.0040737	.1040737	245.4766986	9.6085749	34
35	28.1024368	.0355841	.0036897	.1036897	271.0243685	9.6441590	35
36	30.9126805	.0323492	.0033431	.1033431	299.1268053	9.6765082	36
37	34.0039486	.0294083	.0030299	.1030299	330.0394859	9.7059165	37
38	37.4043434	.0267349	.0027469	.1027469	364.0434344	9.7326514	38
39	41.1447778	.0243044	.0024910	.1024910	401.4477779	9.7569558	39
40	45.2592556	.0220949	.0022594	.1022594	442.5925557	9.7790507	40

NOTE— **N IS EXPONENT N

In interest tables, the time periods usually are expressed in years but can be taken as quarters, months, or any other units of time. A 1-percent interest rate compounded monthly, is *roughly* equivalent to a 12.0 percent rate compounded annually, a 6-percent rate compounded semiannually, or a 3-percent rate compounded quarterly.

To obtain values for fractional time periods or for interest rates not included in an interest table, linear interpolation is usually adequate; if more precision is required, a preprogrammed calculator, a programmable calculator, or a computer can give the answer. A slide rule or a table of logarithems can also be employed.

THE FREQUENCY OF COMPOUNDING

The frequency of compounding has a significant effect on the true interest rate. A carrying charge of 1.0 percent per month on an unpaid department store balance, for example, amounts to an annual rate of about 12.7 percent:

$$(1 + i)^n = (1.01)^{12} = 1.127 = 12.7\%$$

Similarly, a 1.5 percent monthly rate is equivalent to 19.7 percent paid once annually.

The expression $(1 + i)^n$ appears in all six of the basic interest formulas. If the total elapsed time is held constant (say at 1 year) and the compounding period is reduced (or stated another way, the frequency of compounding is increased), the value of this expression increases. As the frequency of compounding approaches infinity, the expression $(1 + i)^n$ approaches e^{in}, indicating "continuous" compounding (where $e = 2.7183 \ldots$). Alternative investments all paying 6.0 percent annually actually yield different amounts, depending on their compounding periods. The effective interest rate on money compounded annually is 6.00 percent; semiannually, 6.09 percent; quarterly, 6.14 percent; bimonthly, 6.15 percent; monthly, 6.18 percent, and continuously, 6.19 percent.

SUMMARY

Lenders charge interest for the use of their money. The financial marketplaces, for either invested or borrowed funds, include capital (or long-term) markets and money (or short-term) markets. The capital market is made up primarily of bonds, corporate equities, and mortgages. The money market includes commercial bank loans, commercial paper, bankers' acceptances, and Small Business Investment Companies. Each type of financial obligation carries its own interest rate, determined by supply-demand relationships. The level of interest rates has a significant impact on the nation's economy, with changes in interest rates causing money to shift from one financial market to another. The most important factor from a business viewpoint is the ease with which long-term capital projects can be financed.

The amount of interest associated with any type of financial transaction can be calculated by using one of six standard interest formulas. These formulas are used in computing: (1) the compound amount of a single payment, (2) the present value of a future payment, (3) the compound amount of an annuity, (4) the sinking fund factor, (5) the capital recovery factor, and (6) the present value of an annuity. In addition to the base interest rate, the frequency with which interest is compounded also has an important influence on the total interest charges associated with an investment.

3

Methods for Evaluating Investment Alternatives

There is a tide in the affairs of men,
Which, taken at the flood, leads on to fortune;
Omitted, all the voyage of their life is bound in
   shallows and in miseries.
On such a full sea are we now afloat;
And we must take the current when it serves,
Or lose our ventures.

William Shakespeare

A wide array of methods exists for evaluating capital expenditures. However, the only effective way to manage investments is to think through carefully the economics of each investment proposal and to make decisions accordingly. Intuition and guesswork, perhaps once adequate as the basis for making capital expenditure decisions, no longer meet the pressures of competition. Making a profit from an investment has become a race against time.

There is only a limited time during which a product or process can remain profitable; maturity and declining profits are a part of every product's life cycle.

To maintain this critical period of profitable operation for as long as possible, investments and costs must be carefully controlled. With the growing trend toward automatic controls—resulting in larger capital investments—depreciation and other fixed charges against capital assets are often more significant than the direct out-of-pocket costs of labor and materials.

It is obvious that effective means of managing capital expenditures are required for financial success in any business venture. Management must avoid passing up profitable investments and making unprofitable ones.

To be of maximum value, the engineer's analysis of investment feasibility must be objective, realistic, easily understandable to his client or his management, and appropriate for the situation. Often, the decision to commit large sums of money may be at stake, and the responsible engineer owes to his sponsor a sound knowledge of the different methods of analyzing investment proposals. Only by being knowledgeable in the tools of investment analysis can he select the approach best suited to the situation.

Basically, an investment decision can be justified by its:

1.   Degree of necessity.
2.   Payout time.
3.   Rate of return.

Payout time is undoubtedly the most widely used measure of the worth of an investment, although rate of return is the most logical and theoretically acceptable means of determining investment feasibility. Degree of necessity, however, still plays an important role in appropriating capital.

INVESTMENTS BASED ON NECESSITY

Capital investments based on necessity include those that an investor would rather not make but must. Typically contained in this unpleasant category are:

1.   Investments on equipment that breaks down and must be replaced.
2.   Investments in capital assets required to meet governmental regulations, such as facilities for air and water pollution control.

3.   Investments required simply to remain in business, such as the replacement of a disaster-stricken facility.

4.   Investments that, even though relatively unprofitable, are necessary to meet competition or retain customers.

Although investments based on necessity are not necessarily profitable, they are usually more lucrative than any available alternatives, since the alternatives are apt to be extreme—such as shutting down altogether. The main difficulty encountered in appraising these investments is the lack of any real objective means of measuring necessity in absolute terms. That lack, however, does not excuse the engineer from using an interest rate in the calculations that select the least undesirable alternative.

PAYOUT TIME IS SIMPLE, FAST, AND MISLEADING

The payout period—the most used measure of investment feasibility—has as its main virtue, simplicity. Payout time is the number of years required for cash earnings or savings generated by a proposed project to equal the original capital investment. As such, the payout time serves as a simple, rough, and readily understood index of project desirability.

Payout time often is used as a screening device to identify projects that are apt to be either exceptionally profitable or unprofitable during their early years. For most purposes, payout time does not provide an adequate measure of investment worth, since many important elements of profitability are not considered. Payback, as a measure of investment desirability, has three important shortcomings:

1.   It overemphasizes the importance of early cash returns in the capital expenditure program.
2.   It ignores the project's economic life.
3.   It fails to consider project earnings after the initial investment has been recovered.

In spite of these shortcomings, the conceptual simplicity of the payout method has certain merits in capital expenditure decisions and actually may provide an adequate measure of investment worth in some situations.

Consider for example, the investment proposal for project A shown in Table 3.1.

Table 3.1 Financial Summary of Proposed Project

Year	(1) Cash Outlay	(2) Net Income before Taxes and Depreciation	(3) Depreciation Charges (120,000/5)	(4) Net Taxable Income	(5) Income Taxes (0.5(4))	Net Income after Taxes (4)-(5)	Net Cash Flow (2)-(5)
0	(120,000)	0	0	0	0	0	(120,000)
1	0	30,000	24,000	6,000	3,000	3,000	27,000
2	0	30,000	24,000	6,000	3,000	3,000	27,000
3	0	40,000	24,000	16,000	8,000	8,000	32,000
4	0	50,000	24,000	26,000	13,000	13,000	37,000
5	0	50,000	24,000	26,000	13,000	13,000	37,000
6	0	50,000		50,000	25,000	25,000	25,000
7	0	50,000		50,000	25,000	25,000	25,000
8	0	50,000		50,000	25,000	25,000	25,000
	(120,000)	350,000	120,000	230,000	115,000	115,000	235,000
							−120,000
							115,000

The investment being considered requires an initial expenditure of $120,-000. The project's estimated life is 8 years, and no salvage value is anticipated. The property is to be fully depreciated over the 5-year period on a straight-line basis.

Income resulting from the investment is estimated at $30,000 for the first year, $30,000 for the second year, $40,000 for the third year, and $50,000 annually for the remaining 5 years. Cash flow, the total amount of money generated by the investment, is found by adding the annual depreciation charge to the after-tax profit.

The payout period answers the single question of how soon the $120,000 investment will be returned from the cash flow; it does not distinguish between profit and depreciation, nor does it consider what happens after the initial investment has been recovered.

The cumulative cash flow by years for this project is as summarized in Table 3.2. The results are plotted graphically in Figure 3.1.

Table 3.2 also shows the discounted value (or present worth) of the cash flows, assuming a 10-percent discount factor. When the interest is ignored, the project pays out some time between the third and fourth year. By interpolation, the payoff period can be estimated at 3.9 years (or determined graphically) in this situation.

When the time value of money is taken into account, the payout time is longer. With a 10-percent discount rate, the payoff period is extended to about 5 years.

Table 3.2 Cash Flow Summary (Project A)

Year	Net Cash Flow (NCF)	Cumulative Cash Flow	Net Cash Flow Discounted at 10% (DCF)	Cumulative DCF
0	(120,000)	(120,000)	(120,000)	(120,000)
1	27,000	(93,000)	24,500	(94,500)
2	27,000	(66,000)	22,300	(72,200)
3	32,000	(34,000)	24,000	(48,200)
4	37,000	3,000	25,300	(22,900)
5	37,000	40,000	23,000	100
6	25,000	65,000	14,100	14,200
7	25,000	90,000	12,800	27,000
8	25,000	115,000	11,700	38,700

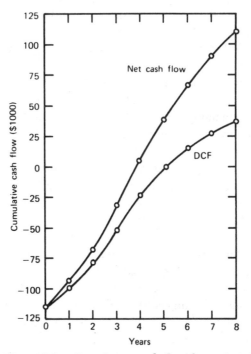

Figure 3.1. *Cumulative cash flow for project A.*

Occasionally, the reciprocal of the payout time is used as another measure of investment desirability, representing the average annual cash recovery during the payout period. The 3.9-year payout indicates an average annual cash return of 1/3.9 or 26 percent of the initial investment. Similarly, the 5-year payout period yields an average cash return of 20 percent annually.

A major disadvantage of the payout period as a meaningful index of investment desirability is evident when another investment proposal is considered, similar in all respects to the previous example, but having a longer life.

This new project (project B, summarized in Table 3.3) requires the same cash outlay as project A and offers the same net income before taxes and depreciation over its first 8 years. Project B, though, is expected to have an economic life of 12 years, so it will produce a $50,000 annual income for an additional 4 years beyond the point where project A terminates. Depreciation is still computed on the same straight-line basis over 5 years.

Project B, during its 12-year life, will generate nearly twice the net cash flow of project A with its 8-year life—$215,000 compared to just $110,000.

Table 3.3 Financial Summary of Project B

Year	(1) Cash Outlay	(2) Net Income before Taxes and Depreciation	(3) Depreciation Charges (120,000/5)	(4) Net Taxable Income (2)−(3)	(5) Income Taxes (0.5(4))	Net Income after Taxes (4)−(5)	Net Cash Flow (2)−(5)
0	(120,000)	0	0	0	0	0	(120,000)
1	0	30,000	24,000	6,000	3,000	3,000	27,000
2	0	30,000	24,000	6,000	3,000	3,000	27,000
3	0	40,000	24,000	16,000	8,000	8,000	32,000
4	0	50,000	24,000	26,000	13,000	13,000	37,000
5	0	50,000	24,000	26,000	13,000	13,000	37,000
6	0	50,000		50,000	25,000	25,000	25,000
7	0	50,000		50,000	25,000	25,000	25,000
8	0	50,000		50,000	25,000	25,000	25,000
9	0	50,000		50,000	25,000	25,000	25,000
10	0	50,000		50,000	25,000	25,000	25,000
11	0	50,000		50,000	25,000	25,000	25,000
12	0	50,000		50,000	25,000	25,000	25,000
	(120,000)	550,000	120,000	430,000	215,000	115,000	335,000
							−120,000
							215,000

Table 3.4   Cash Flow Summary (Project B)

Year	Net Cash Flow (NCF)	Cumulative Cash Flow	Net Cash Flow Discounted at 10% (DCF)	Cumulative DCF
0	(120,000)	(120,000)	(120,000)	(120,000)
1	27,000	(93,000)	24,500	(94,500)
2	27,000	(66,000)	22,300	(72,200)
3	32,000	(34,000)	24,000	(48,200)
4	37,000	3,000	25,300	(22,900)
5	37,000	40,000	23,000	100
6	25,000	65,000	14,100	14,200
7	25,000	90,000	12,800	27,000
8	25,000	115,000	11,700	38,700
9	25,000	140,000	10,600	49,300
10	25,000	165,000	9,600	58,900
11	25,000	190,000	8,000	67,700
12	25,000	215,000	8,000	75,700

As project B's cash flow summary in Table 3.4 (shown graphically in Figure 3.2) illustrates, however, the payout period will be the same as that for project A. The obvious economic advantage of project B over project A is its ability to generate $25,000 for each of 4 years beyond the economic life of A.

In summary, the payout period is often used, and is referred to as a "quick-and-dirty" means of looking at the relative attractiveness of investment proposals. By itself, the payout time is *so* dirty that the analyst cannot see what lies underneath—or worse still, sees the wrong thing. Therefore, the payout period should never be used as the sole criterion in investment evaluation and can seldom be justified for any use other than as an interesting, if irrelevant, piece of information.

THE RATE OF RETURN CONCEPT IS COMPLEX BUT MORE REALISTIC

The rate of return approaches measure the economic worth of an investment by relating the project's anticipated earnings to the amount of capital

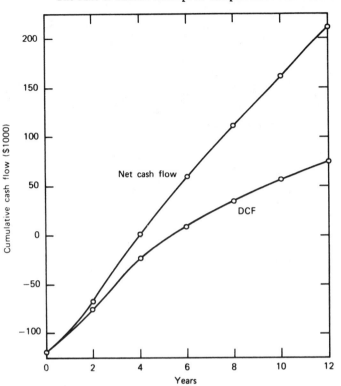

Figure 3.2.   *Cumulative cash flow for project B.*

tied up during the project's estimated life. The main disadvantage of this approach is its relative complexity, but this disadvantage is usually outweighed by the increased precision, thoroughness, and objectivity afforded by the method.

A major difficulty in employing the rate of return technique is the variety of different methods used in calculating the return. There are at least four approaches commonly used in determining the rate of return of a capital investment in a project, each having variations of its own. The four most widely employed methods are:

1. The accounting approach.
2. The operating return approach.
3. The present worth approach.
4. The discounted cash flow (DCF) approach.

THE ACCOUNTING APPROACH.   In the conventional accounting approach to rate of return calculations, the net profit after depreciation and taxes is related to the total or average outstanding investment. This approach assumes that capital recovered through deductions for depreciation against gross income becomes available for use in other projects and should no longer be credited to or charged against the original project.

In referring again to project A, described in Table 3.1, the net profit after taxes over the project's 8-year economic life totals $115,000, an average of about $14,000 per year. Relating this average net profit to the average outstanding investment of $60,000 results in a return on average investment of $14,000/60,000, or 23 percent annually. If, as is sometimes done, the earnings were related to the total initial capital outlay instead of to the average investment, the indicated return would be cut in half, to 11.5 percent.

Similarly, project B (summarized in Table 3.3) nets an average after-tax profit of $215,000/12, or $17,900 annually. This represents an annual return on average investment of 29.9 percent, or a return on total investment of 15 percent.

The principal shortcoming of these accounting procedures is that the effects of compound interest are not considered. This drawback may result in either passing up desirable projects or unknowingly entering into ventures offering low returns. Any time the income pattern, capital outlay schedule, or project life is likely to vary, the rate of return should somehow recognize the timing, as well as the magnitude, of cash expenditures and receipts.

THE OPERATING RETURN APPROACH.   This method expresses the rate of return as the ratio of the average annual cash return to the original investment. The operating return, then, is a measure similar in principle to the accounting method, and as such is subject to many of the same limitations. The operating return approach is used primarily in measuring the relative operating effectiveness of different parts of a business in terms of each operation's gross cash contributions to the business as a whole. However, the measure should be recognized as one of operating efficiency, *not* of economic value.

The operating return can be measured in several different ways, relating either gross operating profits (before depreciation and taxes) or cash flow to either the initial or average investment.

For the project described in Table 3.1 (project A), the average cash inflow over the 8-year period is $29,400 per year ($235,000/8). The operating rate of return on the original $120,000 investment is 24.5 percent, and the

Table 3.5   Present Worth of Net Cash Flow Decreases with Rising Discount
Rates (Project A)

| Year | Net Cash Flow | Present Worth of Net Cash Flow | | | |
		5% Rate	10% Rate	15% Rate	20% Rate
0	(120,000)	(120,000)	(120,000)	(120,000)	(120,000)
1	27,000	25,700	24,500	23,500	22,500
2	27,000	24,500	22,300	20,400	18,700
3	32,000	27,600	24,000	21,100	18,500
4	37,000	30,500	25,300	21,200	17,800
5	37,000	29,000	23,000	18,400	14,900
6	25,000	18,700	14,100	10,800	8,400
7	25,000	17,800	12,800	9,400	7,000
8	25,000	16,900	11,700	8,200	5,800
Total in	235,000	190,700	157,700	133,000	113,600
Total out	120,000	120,000	120,000	120,000	120,000
Net	115,000	70,700	37,700	13,000	(6,400)

return on the average $60,000 investment is 49 percent. Using the average
operating profit of $43,800 ($350,000/8), instead of the cash flow, gives
comparable returns of 36.4 and 72.8 percent, depending on whether the
total or average investment is used as a base.

For project B (Table 3.3) the average cash inflow of $335,000/12, or
$27,900, results in an apparent return of 23.3 percent on the initial invest-
ment or 46.5 percent on the average investment. Similarly, the average
gross operating profit ($45,800) indicates returns of 38.2 percent on the
original investment and 76.4 on the average investment.

As with all other approaches to investment evaluation described thus far,
the operating return overlooks the effects of compound interest. For this
reason, it should never be used when a true measure of profitability is
needed.

THE PRESENT WORTH METHOD.   The present worth approach focuses on
the overall cash consequences of an investment, considering the amount
and timing of cash inflows and outflows and interest rates. The underlying
principle of all present value methods is that money in hand is worth more
than money to be received sometime in the future. The objective of present
value analysis is simply to determine the value today of future cash flows
generated by the investment opportunity over its economic life. This can be
done by selecting an appropriate "cost" or "value" of money and using the

corresponding interest-based discount factors to reduce future amounts to their present worth (as explained in the preceding chapter).

If an investor can obtain 10 percent interest on a deposit, $32,500 to be received in 4 years is worth only $22,200 now, as indicated in Table 3.5. This table also shows that the total present value of all the cash inflows for project A, discounted at 10 percent, is $157,700, or $37,700 more than the investment of $120,000 made at the beginning of the project. If the company's minimum acceptable return in this situation were 10 percent, it would appear desirable to commit $120,000 to this project. With money worth 10 percent, the investor should be willing to pay $157,700 for the op-

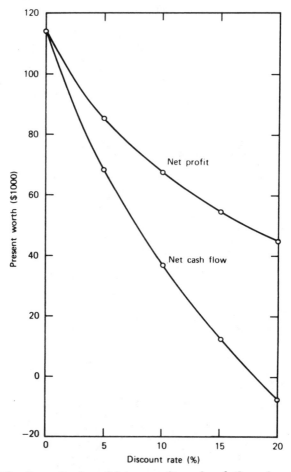

Figure 3.3. *Present value of future profits and cash flows for project A.*

portunity to generate these cash returns; and since only $120,000 is required, it represents a profitable opportunity.

In Table 3.5 the net cash flows for project A are discounted at several different rates. In this situation, in which cash outflows occur early in the project, raising the discount rate decreases the present worth of the project. Project A's discounted net worth is depicted graphically in Figure 3.3, beginning with a net worth of $115,000 on a net (or nondiscounted) basis and dropping to −$6400 to a 20-percent discount rate.

As an alternative to discounting the project's cash flows, the net income after taxes is sometimes discounted. This is shown in Table 3.6, and the relationship between the present worth of discounted net profits and the discount rate is illustrated in Figure 3.3.

Taking the ratio of discounted earnings to the initial investment results in a *benefit/cost ratio*, a term frequently used in connection with some government projects. For tax-free public projects, in which this approach is most often employed, using the discounted cash flows gives essentially the same ratio.

The same information given for project A is summarized for project B in Tables 3.7 and 3.8 and Figure 3.4. Although the same general relationships hold, project B's longer economic life results in a higher present worth, at any given discount rate, than that of project A. Project B's present worth ranges from $215,000 on a nondiscounted basis to $8700 at a 20-percent rate. The present worth of its net income goes from $215,000 to $56,900 over the same range of interest rates.

Even though calculations for the present worth method are straightforward and result in a valid measure of investment worth, there are two main problems associated with this approach:

1. Agreement on what constitutes an appropriate discount rate or minimum acceptable return on investment is sometimes difficult.

2. The answer is expressed in dollars, rather than as a percentage return on invested capital; thus, comparisons among projects having different investment requirements may be awkward.

THE INTERNAL RATE OF RETURN (IRR).   The IRR approach is a special case of the present worth method, in which whatever discount rate applied to the cash flows makes their net discounted values total zero is defined as the DCF rate of return. As might be expected, solving for the appropriate discount rate is mathematically complex (if not impossible), so most analysts

Table 3.6   Present Worth of Net Income After Taxes at Various Discount Rates (Project A)

Year	Net Profit After Taxes	Present Worth of Net Income			
		5% Rate	10% Rate	15% Rate	20% Rate
1	3,000	2,900	2,700	2,600	2,500
2	3,000	2,700	2,500	2,300	2,100
3	8,000	6,900	6,000	5,300	4,600
4	13,000	10,700	8,900	7,400	6,300
5	13,000	10,200	8,100	6,500	5,200
6	25,000	18,700	14,100	10,800	8,400
7	25,000	17,800	12,800	9,400	7,000
8	25,000	16,900	11,700	8,200	5,800
Total	115,000	86,800	66,800	52,500	41,900

Table 3.7   Present Worth of Net Cash Flow at Various Discount Rates (Project B)

Year	Net Cash Flow	Present Worth of Net Cash Flow			
		5% Rate	10% Rate	15% Rate	20% Rate
0	(120,000)	(120,000)	(120,000)	(120,000)	(120,000)
1	27,000	25,700	24,500	23,500	22,500
2	27,000	24,500	22,300	20,400	18,800
3	32,000	27,600	24,000	21,100	18,500
4	37,000	30,400	25,300	21,100	17,800
5	37,000	29,000	23,000	18,400	14,900
6	25,000	18,700	14,100	10,800	8,400
7	25,000	17,800	12,800	9,400	7,000
8	25,000	16,900	11,700	8,200	5,800
9	25,000	16,100	10,600	7,100	4,800
10	25,000	15,300	9,600	6,200	4,000
11	25,000	14,600	8,800	5,400	3,400
12	25,000	13,900	8,000	4,700	2,800
Total in	335,000	250,500	194,700	156,200	128,700
Total out	−120,000	−120,000	−120,000	−120,000	−120,000
Net	215,000	130,500	74,700	36,200	8,700

Table 3.8   Present Worth of Net Income After Taxes at Various Discount Rates
(Project B)

Year	Net Income After Taxes	Present Worth of Net Profit			
		5% Rate	10% Rate	15% Rate	20% Rate
1	3,000	2,900	2,700	2,600	2,500
2	3,000	2,700	2,500	2,300	2,100
3	8,000	6,900	6,000	5,300	4,600
4	13,000	10,700	8,900	7,400	6,300
5	13,000	10,200	8,100	6,500	5,200
6	25,000	18,700	14,100	10,800	8,400
7	25,000	17,800	12,800	9,400	7,000
8	25,000	16,900	11,700	8,200	5,800
9	25,000	16,100	10,600	7,100	4,800
10	25,000	15,300	9,600	6,200	4,000
11	25,000	14,600	8,800	5,400	3,400
12	25,000	13,900	8,000	4,700	2,800
Total	215,000	150,300	103,800	75,900	56,900

employ either trial-and-error or graphical techniques in solving DCF problems. Most computer programs work the same way, but much faster.

The rate of return calculated by using the DCF technique represents continuously compounded interest on invested capital. It assumes that earnings generated by the capital investment are reinvested in the project to continue earning at the same rate. Similar to the present worth method, the IRR approach is concerned with both the magnitude and timing of all cash outlays and receipts. Additionally, it avoids the two main problems inherent in the present worth method by eliminating the need for selecting a sometimes-arbitrary interest rate and by expressing the results as a percentage.

The mechanics of the IRR approach can best be visualized by referring to Table 3.5, in which project A's cash flows are shown discounted at different rates. The present worth of the project's net cash flows is $13,000 at a 15-percent discount rate and drops to $6400 at a 20-percent rate. Since the IRR is defined as the discount rate at which the project's net present worth is zero, the answer obviously lies between 15 percent and 20 percent. Similarly, by referring to Table 3.7 and Figure 3.3. the IRR on project B is found either graphically or by interpolation to be about 22 percent. Computer-facilitated interpolation can be used to estimate the IRR more precisely.

The return on investment found in this manner may be known as the in-

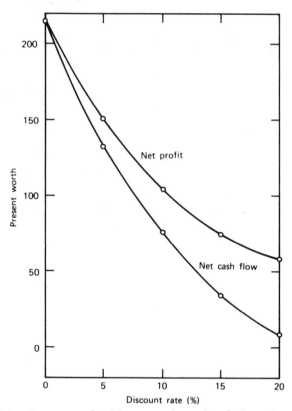

Figure 3.4. *Present worth of future profits and cash flows for project B.*

ternal rate of return, the investor's rate of return, the profitability index, or the discounted cash flow. Whatever it is called, if the indicated rate offers an acceptable return to the investor, the investment proposal presumably will be approved. If the indicated return falls below the investor's minimum acceptable return, the proposal will be rejected. If several alternative investments are available, those showing the highest percentage returns will be given top priority.

The main objections to the IRR approach are:

1. The inference that cash generated by the investment can be reinvested in the project, or at least will continue to earn at the same rate calculated for the project under consideration.
2. The complexity of the calculations involved in computing the rate of return on invested capital.

The first objection is not necessarily valid, however, since the primary purpose of the IRR analysis is simply to reflect the relative productiveness of the capital being committed to the project under consideration, *not* to provide an absolute measure of profitability. The resulting index of capital productivity does provide a valid, objective basis for comparing alternative uses of capital.

The second objection—complexity—can be overcome by using the many available computer programs and short-cut methods designed for solving IRR problems.

SUMMARY

The attractiveness of a proposed investment depends primarily upon the value the investor places on his money and the manner in which he defines

Table 3.9   Comparison of Different Investment Evaluation Techniques

Method of Evaluation	Project A (Table 3.2)	Project B (Table 3.4)
Payout time		
Net	3.9 years	3.9 years
Discounted at 10%	5.0 years	5.0 years
Average annual payout		
Net	26.0%	26.0%
Discounted at 10%	20.0%	20.0%
Accounting return		
Average net profit on original investment	11.5%	15.0%
Average net profit on average investment	23.0%	29.9%
Operating return		
Average operating profit on original investment	36.4%	38.2%
Average operating profit on average investment	72.8%	76.4%
Average cash flow on original investment	24.5%	23.3%
Average cash flow on average investment	49.0%	46.5%
Present worth of net cash flow		
Discounted at 10%	$37,700	$74,700
Discounted at 20%	($6,400)	($8,700)
Present worth of net income		
Discounted at 10%	$66,800	$103,800
Discounted at 20%	$41,900	$56,900
Discounted cash flow return	18.0%	22.0%

that value. Table 3.9 summarizes the results of the different methods of investment evaluation described in this chapter, as applied to two different projects. Either project may appear preferable, depending on which evaluation technique is employed. The wide range of values shown in Table 3.9 emphasizes the necessity for defining the approach to be used and the criteria to be employed prior to conducting an investment analysis. Of this multitude of approaches, the present value and the IRR techniques offer the only valid measures of investment profitability since they, unlike other methods, consider both the amount and timing of all cash inflows and outflows. The IRR expresses profitability in a single percentage figure, which is applicable regardless of the investor's cost of money. Any approach that fails to consider the time value of money cannot be considered a true measure of investment worth.

4

Cash Flow Analysis

To those in the know,
the game is cash flow.

Robert Ringer

The preceding chapter discussed the various ways in which an investment proposal might be evaluated. It was concluded that only approaches based on the interest-based calculation of the present value of future cash flows afford a realistic measure of investment desirability.

This chapter deals with the analysis of cash flows by employing the DCF technique in calculating internal rate of return (IRR). The IRR thus determined serves as a useful and objective measure of the merits of an individual investment or of the relative desirability of several alternative investments. The IRR, as an analytical measure of investment worth, provides management with a sound basis for approving (or disapproving) an investment proposal, or for choosing the best of several opportunities.

IRR CALCULATIONS BASED ON THE DCF CONCEPT

Making a profitable investment is not especially difficult once the investment opportunity has been pointed out; but identifying a profitable investment opportunity, or the best of several opportunities, requires a considerable amount of mathematical prophecy.

Most engineering projects involve a choice between alternatives having different investment requirements and perhaps covering different time periods. To select the best possible alternative, it is important to have an objective means of evaluating all alternatives on a consistent and comparable basis.

The DCF approach is generally recognized as a logical, and by many as the *most* logical, way to analyze and to compare alternative investment opportunities involving different time periods and generating different cash (or benefit) flows.

The term cash flow, as used here, is defined as the net profit after taxes, plus depreciation, depletion, and other "paper" expenses that were deducted from gross income. Cash flow represents the total amount of money generated by the project and available for other uses. Cash outflows are actual out-of-pocket expenditures; cash inflows are similar to money in the bank. The net cash flow, then, is simply the difference between the cash inflows and outflows. Cash flow is essentially a function of price, cost, volume, investment requirement, depreciation, and tax structure.

Net cash flow is *not* the same as profit. Cash flow represents only the difference between total cash receipts from the business's operations and the total cash outlays incurred in carrying out these operations. Depreciation, which must be considered in calculating profit, is not a cash expense and therefore affects cash flow only by reducing the out-of-pocket income tax liability.

In DCF analysis, a project's net cash flow is estimated for each year of its projected economic life. Then these cash flows are discounted at whatever interest rate makes the sum of discounted cash inflows equal to the sum of discounted cash outflows.

The interest rate that results in the discounted cash inflows being equal to the discounted cash outflows represents the project's IRR, also referred to as its return on investment (ROI) or its DCF rate of return. Whatever its name, this interest rate represents the highest rate at which the required capital can be borrowed and paid off from the cash earnings generated by the project over its economic life. If money were available at a lower interest rate, the debt could be retired with something left over; if the com-

Table 4.1   Effect of Cash Flow Timing on Project IRR

Year	Net Cash Flow		
	Project 1	Project 2	Project 3
0	−$1,000	−$1,000	−$1,000
1	300	500	100
2	300	400	200
3	300	300	300
4	300	200	400
5	300	100	500
IRR	15%	20%	12%

pany's cost of capital were higher than the indicated rate, the cash generated by the project would not be enough to pay off the borrowed money.

The following example illustrates the effect of timing of cash flows on a project's IRR, as measured by the DCF approach. Table 4.1 shows three simple 5-year projects, each requiring an initial cash outflow of $1000 and each followed by 5 years of cash inflows totaling $1500. The first case offers a cash return of $300 annually; the second has a declining pattern of cash flows, starting at $500 and dropping by $100 each year; and the third has a first-year cash return of $100, which increases by $100 each year to $500 in the fifth year.

Although the net totals of cash inflows and outflows are the same in each case—$1000 out and $1500 in—there is considerable difference in the relative economic desirability of these projects. Obviously, project 2 is preferred, because it returns the investment most quickly, whereas project 3 appears least desirable for the opposite reason. As measured by the DCF technique, project 1 has a 15-percent IRR; project 2, a 20-percent IRR; and project 3 yields a 12-percent return. Thus, the DCF approach objectively puts the projects in the same order as is intuitively obvious. This is perhaps one of the fundamental benefits of DCF analysis—especially when the results are less obvious.

PROBLEMS IN DCF ANALYSIS

A chief drawback to using the DCF method of analysis is the difficulty involved in finding the appropriate interest rate; unless a computer program is available, many tedious trial-and-error calculations may be necessary. For example, consider the cash flow schedules described in Table 4.1.

The conventional trial-and-error approach to solving these problems is to first total the net cash inflows and outflows. The excess of total inflows over total outflows indicates that the projects are earning some return on the initial investment.

Next, some interest rate (e.g., 10 percent) is selected; each annual net cash flow is discounted by a factor appropriate to that rate, and the discounted inflows are compared with the discounted outflows. At a 10-percent rate, discounted inflows are still higher than discounted outflows in each case, indicating that the projects are all earning at a rate higher than 10 percent.

The same process is then repeated, using perhaps a 20-percent rate for this trial. When discounted cash inflows are compared with discounted cash outflows, the cash flows for project 2 total zero, indicating a 20-percent IRR. For the other two projects, the discounted cash outflows exceed the inflows, indicating IRRs between 10 and 20 percent.

For projects 1 and 3, the next step could employ interpolation between 10 and 20 percent to approximate their IRRS; or other rates between 10 and 20 percent could be tried, until the correct rates are found.

In any event, the arithmetic involved in DCF analysis can be tedious and is undoubtedly one of its chief drawbacks.

DOUBLE AND IMAGINARY SOLUTIONS FROM SIGN CHANGES

Because of the higher-order equations involved in complex DCF analyses, it is sometimes possible for a single problem to have several solutions. The misleading results that are possible from a DCF problem involving multiple solutions are illustrated in the following situation.

Year	Net Cash Flow ($)
0	+1000
1	−3000
2	+2000

On visual examination, the merits of such an investment appear highly questionable; with the total cash inflow equaling the total cash outflow, a zero-percent IRR obviously applies. This problem can be solved mathematically for the IRR by setting the sum of the DCFs equal to zero and solving for the discount rate R:

$$1000 - \frac{3000}{(1 + R)} + \frac{2000}{(1 + R)^2} = 0$$

Solving for R:

$$1000(1 + R)^2 - 3000(1 + R) + 2000 = 0$$
$$R^2 - R = 0$$
$$R(R - 1) = 0$$
$$R = 0; R = 1$$

The IRR in this case can be either zero or 100 percent, both of which satisfy the DCF definition of IRR:

Year	Net Cash Flow ($)	Net Cash Flow Discounted at 0%	Net Cash Flow Discounted at 100%
0	+1000	+1000	+1000
1	−3000	−3000	−1500
2	+2000	+2000	+ 500
		0	0

Upon examination of the problem, the 100-percent IRR figure becomes suspect. Although mathematically correct, it is obviously unrealistic from an investment standpoint.

In some problems, the presence, let alone the practicality, of a double solution may be far less apparent. Any trial-and-error solution, whether obtained manually or on a computer, may find one mathematically correct answer but ignore the other.

Fortunately, the possibility of such situations occurring can be easily identified by examining the cumulative cash flow pattern. A multiple solution can occur each time the cumulative cash flow changes sign. Such changes do not happen in most projects, because there is generally a substantial cash outflow during the initial time period, followed by a series of cash inflows over the remainder of the project's economic life. Should heavy expenditures be required toward the middle or end of the project's life, however, the possibility of a double solution exists. Secondary recovery of oil is a good example, in which substantial expenditures may be required to restimulate a declining oil flow—essentially a "double investment" situation.

Imaginary solutions to DCF problems may also occur in the same type of situations that lead to double solutions. A net cash flow of −1000, +2000, −2000, for example, yields an imaginary result. Such a cash flow pattern, however, is so clearly unattractive as to discourage further consideration.

MATHEMATICAL SOLUTION OF DCF PROBLEMS

The complexity of most DCF problems precludes any direct mathematical solution. Only when a 2- or 3-year economic life is assumed, or when there is a uniform series of cash flows, can a solvable equation be formulated. The general form of a DCF problem can be expressed as:

$$\text{NCF}_0 + \frac{\text{NCF}_1}{(1 + R)} + \frac{\text{NCF}_2}{(1 + R)^2} + \frac{\text{NCF}_3}{(1 + R)^3} + \ldots + \text{NCF}_n = 0$$

where NCF_n represents the net cash flow occurring in the nth year. To estimate IRR requires that this equation be solved for R—a task far beyond the abilities of most mathematicians.

The simplifying assumption of a uniform cash flow pattern may be realistic enough for some projects, especially if the object of the analysis is to screen out undesirable projects or to obtain a rough approximation of a project's IRR when insufficient data are available for more thorough analysis. In such a situation, a quick mathematical solution can be quite useful.

When a single cash outflow (representing the initial capital investment) is made at the beginning of a project, followed by uniform cash inflows over the remainder of the project's economic life, the resulting IRR can be found by using the formula for computing the capital recovery factor (see chapter 2).

$$\text{capital recovery factor} = \frac{R(1 + R)^n}{(1 + R)^n - 1} = \frac{\text{annual cash inflow}}{\text{initial investment}}$$

where the initial investment is made in year 0 and the annual cash inflow takes place uniformly over the years 1 through n.

Given the initial cash outflow, the annual cash inflows, and the economic project life, solving the above equation for R gives the IRR.

While the equation is solvable, a good set of interest tables makes the job much easier. The ratio of the annual cash inflow to the initial investment gives the capital recovery factor. All that need be done is to find the interest rate having that capital recovery factor for the desired number of years. For example, if the capital recovery factor were 0.300 for a 6-year period, examination of interest tables would show a 20-percent interest rate to have a capital recovery factor of 0.3007057 for a 6-year uniform annual cash recovery. The IRR in this case, therefore, would be about 20 percent.

This approach, as employed in screening new projects, is discussed in some detail in a later section, where a mathematical approximation is presented.

TRIAL-AND-ERROR SOLUTIONS—WITH A COMPUTER

A mathematical trial-and-error approach can be used to "bracket" a project's IRR. Then the approximate IRR can be calculated by interpolation. This procedure is commonly used and can give a satisfactory solution without a great deal of computation time. A maximum of about five trial solutions is generally sufficient.

A distinct advantage of this approach is that it can be set up in a standard format so that the necessary calculations can be performed in a routine fashion by clerical personnel.

Figure 4.1 shows a typical DCF worksheet. The net cash expenditures for each year of the project's life are inserted in the appropriate column of the upper table, and the net cash receipts in the lower table. Each annual cash flow is then discounted by multiplying it by its corresponding discount factor for the indicated 10-, 15-, 25-, and 40-percent trial interest rates. This is done for both cash outflows and cash inflows. Finally, the columns are totaled, and the ratio of total discounted receipts to total discounted investments is calculated.

Since the return on investment is by definition the interest rate that makes the discounted inflows equal to the discounted outflows, a ratio of discounted receipts to discounted expenditures of 1.000 identifies the interest rate being sought. A ratio of discounted inflows to discounted outflows greater than 1.0 indicates that the trial interest rate is too low; similarly, a ratio of less than 1.0 indicates that the trial rate is too high.

Having computed the ratios for the five trial interest rates—0, 10, 15, 25, and 40 percent—the return on investment percentage can be interpolated by means of the following formula:

$$\text{IRR} = a + (b - a) \frac{a_r - 1.00}{a_r - b_r}$$

where a is the trial interest rate at which the ratio of discounted receipts to discounted expenditures is larger than, but closest to, 1.0; b is the trial interest rate having a ratio smaller than, but closest to, 1.0; and a_r and b_r are their respective ratios.

For example, if the ratio of discounted receipts to discounted expendi-

Year	Trial No. 1 0% Interest Rate Actual Amount	Trial No. 2 10% Interest Rate Discount Factor	Present Value	Trial No. 3 15% Interest Rate Discount Factor	Present Value	Trial No. 4 25% Interest Rate Discount Factor	Present Value	Trial No. 5 40% Interest Rate Discount Factor	Present Value
0		1.000		1.000		1.000		1.000	
1		0.909		0.870		0.800		0.714	
2		0.826		0.756		0.640		0.510	
3		0.751		0.658		0.512		0.364	
4		0.683		0.572		0.410		0.260	
5		0.621		0.497		0.328		0.186	
6		0.564		0.432		0.262		0.133	
7		0.513		0.376		0.210		0.095	
8		0.467		0.327		0.168		0.068	
9		0.424		0.284		0.134		0.048	
10		0.386		0.247		0.107		0.035	
Total cash outflow									
0		1.000		1.000		1.000		1.000	
1		0.909		0.870		0.800		0.714	
2		0.826		0.756		0.640		0.510	
3		0.751		0.658		0.512		0.364	
4		0.683		0.572		0.410		0.260	
5		0.621		0.497		0.328		0.186	
6		0.564		0.432		0.262		0.133	
7		0.513		0.376		0.210		0.095	
8		0.467		0.327		0.168		0.068	
9		0.424		0.284		0.134		0.048	
10		0.386		0.247		0.107		0.035	
Total cash inflow									
Inflow/ outflow ratio									

ROI calculation

Figure 4.1. *Typical DCF worksheet is set up for trial-and-error analysis.*

tures is 1.15 at a 15-percent trial interest rate and 0.80 at a trial rate of 25 percent, the IRR is approximately as follows.

$$\text{IRR} = 15 + (25 - 15)\,\frac{1.15 - 1.00}{1.15 - 0.80} = 19 \text{ percent}$$

The interest factors shown in Figure 4.1 are the standard end-of-year present value rates found in most interest tables, representing the $1/(1+R)^n$ factor (see Chapter 2). Some companies prefer to use an average midyear factor instead; thus, the discount factor applying to investments or receipts incurred during year n is $1/(1+R)^{n-0.5}$. In either case, the approach is the same, and the difference in results is not critical, so long as the preferred method is applied consistently.

GRAPHICAL SOLUTION OF DCF PROBLEMS

The conventional graphical solution to DCF problems simply substitutes a graph for the mathematical interpolation described in the preceding section (see Figures 3.3 and 3.4). Once the ratios of discounted cash inflows to discounted cash outflows have been calculated for the selected trial interest rates, the ratios are plotted on a graph. A smooth curve may then be drawn through the points, and the discount rate corresponding to the point on the curve having a ratio of 1.0 is the IRR.

Figure 4.2 shows how the graphical solution technique works. Here the following values were obtained for the five trial interest rates.

Trial Interest Rate (%)	Ratio of Discounted Cash Inflow to Discounted Cash Outflow
0	2.18
10	1.45
15	1.22
25	0.91
40	0.64

These five points are plotted, as shown in Figure 4.2, with the ratios on the vertical axis and the interest (or discount) rates on the horizontal axis. A horizontal line has been drawn at a ratio of 1.0, and the point at which the curve intersects this line identifies the project's IRR. In the example, the IRR is 21.5 percent.

By contrast, a linear interpolation between the 15- and 25-percent rates, using the equation from the preceding section, gives an answer to this

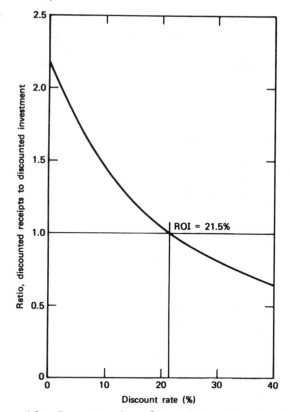

Figure 4.2. *Conventional graphical solution of DCF problem.*

same problem of 22.1 percent. The relationship between discounted cash flow values and interest rates however, is not linear, which means that the graphical method is more accurate.

SCREENING NEW PROJECTS

Complicated problems may require complicated solutions, and complex questions may require complex answers. But complexity should be avoided whenever possible. Complex solutions may guarantee precise answers, but they cannot result in a level of accuracy beyond that of the data employed.

Most companies are interested in pursuing whatever profitable new opportunities may come along. Frequently, there are many different opportunities under consideration at the same time. Although most are probably not worth following, a few may be profitable, and an occasional one may be especially rewarding. How to classify all the available opportunities and to distinguish the most promising is an often perplexing management problem.

For purposes of classification, it is especially important to keep economic studies simple in their early stages. The accuracy of available economic data may preclude any precise evaluation and seldom justifies even a serious attempt at precision.

As early as possible in an economic investigation, simple calculations should be carried out to determine whether the overall concept being considered is worthy of further study. The data can be refined and the details can be filled in later. It is seldom desirable to undertake any elaborate or sophisticated analysis in a project's early stages with the thought of obtaining a complete answer on all points.

The "screening test" is a type of analysis aimed at determining whether a particular proposal is worth pursuing further, or at finding which of several alternatives may offer the best opportunities for further investigation.

A "PROGRAMMED" APPROACH TO PROJECT SCREENING

Table 4.2 offers a programmed approach to screening new projects. The program is written for users of computerized (electronic) spread sheets rather than computer programmers. It requires only six input figures. It results in an approximate IRR percentage, having employed the DCF approach.

The six inputs required in this program and the values used for each in the two examples shown in Table 4.2 are:

Line	Item	Example A	Example B
1	Average annual sales or revenues	$200,000	$200,000
2	Direct production costs	$100,000	$120,000
3	Indirect and overhead costs	$ 40,000	$ 50,000
4	Net investment	$100,000	$ 80,000
5	Economic project life	5 years	8 years
6	Income tax rate	50%	50%

Average annual sales require that an estimate be made of the number of units sold at a given price, less direct selling costs and commissions.

Similarly, *direct production costs* (line 2) refer primarily to the unit costs of labor and materials incurred at the assumed production volume.

Table 4.2 Project Evaluation Program

Line	Item	Units	Source	Example A	Example B
1	Average annual sales	$	Input	200,000	200,000
2	Annual direct production costs	$	Input	100,000	120,000
3	Annual indirect and overhead costs	$	Input	40,000	50,000
4	Initial investment	$	Input	100,000	80,000
5	Economic project life	Years	Input	5	8
6	Income tax rate	%	Input	50	50
7	Average annual depreciation expense	$	Line 4/line 5	20,000	10,000
8	Total annual deductions	$	Line 2 + line 3 + line 7	160,000	180,000
9	Net annual profit before taxes	$	Line 1 − line 8	40,000	20,000
10	Annual income taxes	$	Line 6 × line 9	20,000	10,000
11	Net annual profit after taxes	$	Line 9 − line 10	20,000	10,000
12	Net annual cash flow	$	Line 11 + line 7	40,000	20,000
13	Capital recovery rate (see Chapter 2)	Ratio:	Line 12/line 4	0.400	0.250
14	IRR	$	Graph	29	19

Indirect and overhead costs (line 3) are generally accrued on a time basis rather than on a per-unit-of-production basis; they include capital charges, administrative and general expenses, and other time-related costs of carrying on the proposed venture.

Net investment (line 4) refers to the total capital required to put the venture in the startup stage, less any appropriate tax credits. Working capital requirements need not be considered in this initial screening.

Economic project life (line 5) is number of years during which the assumed conditions will persist and over which the original investment must be recovered. A project life of between 5 and 10 years is commonly used. Less than 5 years is usually unrealistic, and conditions beyond 10 years can seldom be anticipated.

Income tax rate (line 6) refers to the total of all income-based taxes—federal, state, and local—plus any surcharges that might apply.

Lines 7 through 13 in the screening program consist of the calculations necessary to determine the capital recovery rate (line 13), which can be translated directly into a percentage IRR by referring to the graph (Figure 4.3) or to interest tables containing capital recovery factors for a wide range of compound interest values. The graph is entered on the vertical axis at the computed capital recovery rate, and the IRR is read on the horizontal scale for the appropriate project life.

ARITHMETIC APPROXIMATION OF THE IRR

To calculate the *approximate* return on investment without reference to either a graph or interest tables, the following three lines can be added to the program.

Line	Item	Units	Source	Example A	Example B
15	Depreciation rate	Ratio	1.00/line 5	0.200	0.12̇5
16	Excess return rate		Line 13 − line 15	0.200	0.125
17	Approximate internal rate of return	%	150[a] × line 16	30	19

[a] An experience-derived multiplier that gives close approximations to the graphical method.

These approximate IRR figures compare favorably with the values obtained graphically: 30 percent against 29 percent, and 19 percent against 18.6 percent.

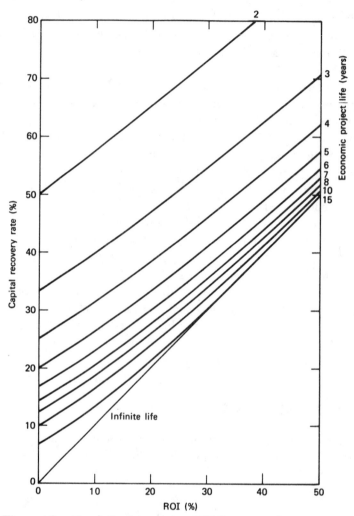

Figure 4.3. *Graph for determining ROI from capital recovery rate.*

ASSUMPTIONS

Two important assumptions are necessary in this highly simplified approach to project evaluation:

1. Annual sales, costs, and tax rates remain constant over the project's economic life.

2. The entire investment is depreciated (or amortized) on a uniform straight-line basis over the project's economic life.

The program presented here gives an accurate solution to the problem, under the conditions imposed by these simplifying assumptions. If the assumptions are tolerable, then the answers will be adequate for their intended purpose.

This is a quick way to get a good "feel" for the economic attractiveness of a proposed project in its early stages.

SENSITIVITY ANALYSIS FOR NEW PROJECTS

Sensitivity can be roughly defined as the change in output brought about by a specified change in input. The main purpose of sensitivity analysis is to determine which of several inputs may be the most important and the extent to which each input affects the output.

The project evaluation program presented in the preceding section offers a convenient means of identifying the sensitivity of a project's IRR to changes in annual revenues, direct and indirect costs, economic life, capital investment requirements, and income tax rates. Thus, the relative importance of different economic parameters can be assessed during the early planning stages of a project, before any major commitments are made. Additional investigation can then be directed toward clarifying the most relevant factors.

Although the program described in Table 4.2 requires six inputs, variations in only three of the six inputs need be considered for most projects. Since revenues, direct costs, and indirect costs are eventually combined, they can be considered together as a single parameter; only the net difference between revenue and cost is significant. In Table 4.3, costs are held constant while the sales revenue is varied to generate the different levels of net operating profit—the net amount of cash left after all operating costs and expenses have been paid, but before deductions for depreciation and taxes have been taken. The same effect could be obtained by holding the sales constant while varying the cost figures.

In looking at new projects, there are many uncertainties involved, and it is often difficult to define all the conditions under which the project must operate. Predesign estimates of capital requirements might be off by as much as 30 percent, and estimates of economic life are always uncertain,

Table 4.3 Effect of Variables on Investment Profitability

Line[a]	Parameter	Base Case	Variations					
			1	2	3	4	5	6
1	Average annual sales	$200,000	$170,000	$230,000	$200,000	$200,000	$200,000	$200,000
2	Annual direct production costs	100,000	100,000	100,000	100,000	100,000	100,000	100,000
3	Annual indirect and overhead costs	40,000	40,000	40,000	40,000	40,000	40,000	40,000
4	Initial investment	100,000	100,000	100,000	50,000	150,000	100,000	100,000
5	Economic project life (years)	8	8	8	8	8	4	12
6	Income tax rate (%)	50	50	50	50	50	50	50
12	Net annual cash flow	$36,000	$21,000	$51,000	$33,000	$39,000	$43,000	$34,000
13	Capital recovery rate	0.36	0.21	0.51	0.66	0.26	0.43	0.34
14	ROI (%)[b]	33	14	50	65	21	25	34

[a] From Table 4.2.
[b] From Figure 4.3.

since they depend upon a number of uncontrollable external and internal factors. Income tax considerations perhaps play as great a role as anything in a company's decision to become involved in a new project; however, for a particular situation, only one tax rate applies, and this rate can usually be identified in advance.

Even with only three combinations of parameters to consider, the problem can become quite complex if several different values are applied to each of the three. Examining just three factors, each having three different values, results in 27 different combinations. It is not difficult to visualize the unmanageable proportions that such a problem could easily reach; this is why a computer is often used to calculate the effect of *all* possible outcomes whenever it is necessary to go into such detail.

Alternatively, the combination of the computer and the Monte Carlo method of randomly selecting levels of the input variables can shorten the time to estimate a realistic range of IRR estimates.

The necessity for looking at all possible combinations of variables can also be minimized by using the best estimate for each parameter to form a base case. One item at a time is then varied with all other parameters held constant. This reduces the number of combinations to just seven—the base case plus six variations, two for each of the three parameters.

The parameters used in the seven different combinations are summarized in Table 4.3 All seven situations assume $100,000 in direct costs, $40,000 in indirect costs, and a 50-percent income tax rate. The base case—representing the best estimate of each parameter—shows a total sales revenue of $200,000, a capital investment of $100,000, and an 8-year economic life. After carrying out the calculations described in the project evaluation program (Table 4.2), the base case is found to have an IRR of 33 percent.

Variations 1 and 2 examine the sensitivity of the IRR to changes in the net operating profit. Variation 1 assumes a 50-percent decrease in the net operating profit, whereas variation 2 assumes a 50-percent increase. (The 50-percent change in net operating profit is brought about by just a $30,000 change in sales revenues.) Their respective returns are 14 and 50 percent.

Variations 3 and 4 assume changes in the capital investment requirement. Variation 3, with a 50-percent lower investment, shows a net return of 65 percent. Variation 4, requiring 50 percent more capital than the base case, has a 20.5-percent IRR.

Variations 5 and 6 decrease and increase the estimate of economic project life by half. The shorter life (4 years) lowers the IRR to 25 percent; the longer life (12 years) raises the return to 34 percent.

Table 4.4   Sensitivity of IRR to Parameter Changes

Variation	Description	IRR	Change from Base Case %
	Base Case	33	
1	50% Decrease in net operating profit	14	−58
2	50% Increase in net operating profit	50	+52
3	50% Decrease in capital investment	65	+97
4	50% Increase in capital investment	21	−36
5	50% Decrease in economic life	25	−24
6	50% Increase in economic life	34	+3

The results of this sensitivity analysis are summarized in Table 4.4, which shows the percentage change in the project's IRR brought about by a 50-percent change in either direction for each of the three parameters being examined.

In the preceding example, errors in forecasting the economic life of a project have relatively little effect on its apparent profitability. A high-side error is far preferable to a low-side error, since it has far less effect on the project's IRR (see variations 5 and 6 in Table 4.3).

The IRR is especially sensitive to variations in capital requirements, with high-side estimates again preferred over estimates on the low side. The same is true to a lesser extent with net operating profit.

Under these conditions, it is better (in terms of overall accuracy) to over-estimate than to underestimate the values of the various parameters involved in a project evaluation.

PROBABILITY AND PROJECT EVALUATION

Another important aspect of the project evaluation problem—one that is particularly significant in the predesign or early design phases of a new or proposed project—is the accuracy of the various estimates used in evaluating the project.

Of the parameters studied in the sensitivity analysis, IRR is least sensitive to changes in a project's economic life, but is greatly affected by changes in the capital investment requirement and in the expected level of net operating profits.

In addition to knowing how much the IRR is affected by a parameter change, it is important to know the likelihood of that change occurring. When the amount and probability of change are known, an expected value can be determined and used as the basis for decision making.

The expected IRR is defined (loosely) as the sum of the products of the IRRs that will result from each possible set of conditions and the probability of each possible set of conditions occurring.

For example, if there is a 20-percent chance that a project will achieve a 30-percent IRR, a 50-percent chance of its earning 20 percent, and a 30-percent chance of realizing a 10-percent return, the expected IRR is:

$$
\begin{array}{rcl}
0.2 \times 30 & = & 6.0 \\
0.5 \times 20 & = & 10.0 \\
0.3 \times 10 & = & \underline{3.0} \\
& & 19.0 \text{ percent}
\end{array}
$$

Computing the expected IRR for a new project, therefore, requires that probabilities be assigned to each possible combination of likely parameters. The result of each combination must then be weighted by the probability of its occurrence. The weighted average of all possible combinations represents the expected IRR.

In the project evaluation program discussed in the last two sections, six parameters were employed in estimating the IRR on a new project:

1. Sales revenues.
2. Direct Costs.
3. Indirect costs.
4. Capital investment.
5. Economic life.
6. Income tax rate.

For convenience, the first three items are combined into a single item–the net operating profit before deductions for depreciation and taxes. By varying just the sales revenue, any desired level of operating profit can be obtained. Also, since only one income tax rate applies to a given project, there is little point in introducing several different values for it.

Three factors remain to vary: operating profit (via changes in the sales revenues), capital investment requirement, and economic project life. If three different values were assigned to each of these three items—a low, median, and high estimate, or an optimistic, pessimistic, and most likely estimate—the resulting 27 different combinations would be as shown in Figure 4.4.

In Figure 4.4, each of the three specified project lives is broken down into three levels of capital investment; each level of capital investment is in turn associated with three different levels of sales revenues.

Determining the internal rate of return associated with each of the 27 combinations shown in Figure 4.4 is a simple enough task if an electronic spread sheet is used for the project. The returns range from a low of −4.0

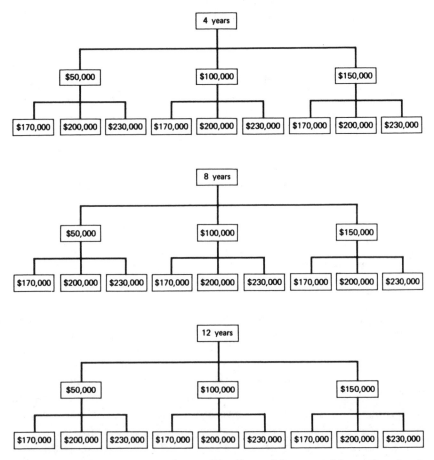

Figure 4.4.   *There are 27 possible combinations of three variables at three levels.*

percent (for a 4-year project having an investment requirements of $150,-000 and generating sales of $170,000 annually) to a high of 96 percent (for an 8-year project with a $50,000 investment and $230,000 in annual sales).

Identifying the probability of occurrence of each possible outcome requires multiplying the probabilities of each of the three elements—economic life, investment, and sales. For example, if the probability of achieving the low, median, and high estimates for the three parameters were as follows:

	Low	Median	High
Economic Life	0.1	0.7	0.2
Capital investment	0.2	0.5	0.3
Sales revenue	0.4	0.4	0.2

Then the probability of a combination consisting of a median economic life, a high capital investment, and a low sales revenue would be $0.7 \times 0.3 \times 0.4 = 0.084$, or 8.4 percent. Similarly, a high economic life, a low capital investment, and a high sales revenue are expected to occur only $0.2 \times 0.2 \times 0.2 = 0.008$, or 0.8 percent of the time. The most likely combination, a median value of each parameter, has a probability of $0.7 \times 0.5 \times 0.4 = 0.140$, or 14.0 percent. The sum of the probabilities of all 27 combinations is 1.00, because it is assumed that all possibilities are covered.

By calculating the probability of occurrence of each of the 27 possible combinations shown in Figure 4.4 and multiplying the IRR associated with each combination by its probability, the expected IRR of the project as a whole can be calculated. This is illustrated in Table 4.5. The expected IRR is about 30 percent.

By approaching the problem in this manner, the chances of achieving any specified IRR, or the probability that the IRR will fall within a given range, can be derived by plotting the cumulative probability distribution of the various outcomes from Table 4.5, as shown in Figure 4.5. Here the probability of attaining at least a specified IRR is plotted from the data in Table 4.5.

To find the probability of earning between 30 and 40 percent from the graph, it is first necessary to find the probabilities of attaining these returns. The graph indicates a 40-percent chance of earning at least 30 percent, and a 22.5-percent chance of reaching 40 percent. Thus, the probability of earning between 30 and 40 percent is 40.0–22.5, or 17.5 percent. Sixty percent of the possible outcomes will fall below 30 percent, and 22.5 per-

Table 4.5 Calculation of Expected IRR for New Project

Economic Life (Years)	Capital Investment ($)	Sales Revenues ($)	IRR ($)	Probability of Occurrence	Expected Value of IRR (%)
4	50,000	170,000	25.0	0.008[a]	0.200
4	50,000	200,000	62.0	0.008	0.496
4	50,000	230,000	95.5	0.004	0.382
4	100,000	170,000	4.0	0.020	0.080
4	100,000	200,000	25.0	0.020	0.500
4	100,000	230,000	44.0	0.010	0.440
4	150,000	170,000	(4.0)	0.012	(0.048)
4		0.012	0.138		
4	150,000	230,000	25.0	0.006	0.150
8	50,000	170,000	32.5	0.056	1.820
8	50,000	200,000	65.0	0.056	3.640
8	50,000	230,000	95.8	0.028	2.682
8	100,000	170,000	13.5	0.140	1.890
8	100,000	200,000	32.5	0.140	4.550
8	100,000	230,000	49.5	0.070	3.465
8	150,000	170,000	6.5	0.084	0.546
8	150,000	200,000	20.5	0.084	1.722
8	150,000	230,000	32.5	0.042	1.365
12	50,000	170,000	33.0	0.016	0.528
12	50,000	200,000	64.2	0.016	1.027
12	50,000	230,000	94.2	0.008	0.754
12	100,000	170,000	16.0	0.040	0.640
12	100,000	200,000	33.0	0.040	1.320
12	100,000	230,000	48.5	0.020	0.970
12	150,000	170,000	9.0	0.024	0.216
12	150,000	200,000	22.0	0.024	0.528
12	150,000	230,000	33.0	0.012	0.396
				Total: 1.000	Total: 30.397

[a] $0.1 \times 0.2 \times 0.4 = 0.008$.

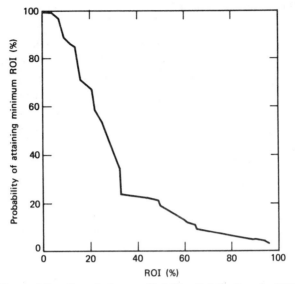

Figure 4.5. *Cumulative profitability distribution for ROIs.*

cent will fall above 40 percent, leaving 17.5 percent in the 30–40 percent range.

To avoid the tedious calculations involved in such an approach, a simpler method can be substituted, with satisfactory results in most cases. Instead of calculating the expected IRR associated with each possible combination of parameters, the IRR associated with the expected parameters can be used as a single index of profitability. Utilizing the example just discussed, the expected economic life can be calculated by multiplying the low, median, and high estimates by their respective probabilities: 0.1×4 years $+ 0.7 \times 8$ years $+ 0.2 \times 12$ years $= 8.4$ years expected life. In the same way, the expected capital investment is $105,000, and the expected operating profit is $194,000. Putting these expected values together (as shown in Table 4.6) results in an expected annual net cash flow of $33,250, representing a capital recovery rate of 0.317. The IRR associated with this capital recovery rate over the expected 8.4-year period is approximately 28 percent, which can be considered the expected profitability of the proposed investment. This compares with the 30-percent return found by the more detailed procedure.

This approach does not identify the range of values possible in a given situation and does not identify the likelihood of achieving various levels of

Table 4.6   Return on Investment Calculations under Selected Conditions

Line	Expected Conditions	Worst Possible Conditions	Most Likely Conditions	Best Possible Conditions
1 Average sales ($)	194,000	170,000	200,000	230,000
2 Direct costs ($)	100,000	100,000	100,000	100,000
3 Indirect costs ($)	40,000	40,000	40,000	40,000
4 Net investment ($)	105,000	150,000	100,000	50,000
5 Economic life (years)	8.4	4	8	8
6 Income tax rate (%)	50	50	50	50
7 Average depreciation ($)	12,500	37,500	12,500	6,250
8 Total deductions ($)	152,500	177,500	152,500	146,250
9 Net profit before taxes ($)	41,500	(7,500)	47,500	83,700
10 Income taxes ($)	20,750	(3,750)	23,750	41,875
11 Net profit after taxes ($)	20,750	(3,750)	23,750	41,875
12 Net cash flow ($)	33,250	33,750	36,250	48,125
13 Capital recovery rate	0.317	0.225	0.363	0.963
14 ROI (%)	28	−4.0[a]	33	96

[a] Any capital recovery rate of less than $1/N$ (where N is the economic life in years) does not recover the entire investment over the project's life and, therefore, represents a negative ROI. A capital recovery rate of $1/N$ corresponds to a zero percent return; any rate of more than $1/N$ indicates a positive return.

profitability. It does, however, give a quick, convenient, and realistic index of project desirability.

ECONOMIC PROJECT LIFE

Most feasibility studies involve a present expenditure made to achieve either earnings or savings over several years in the future. This situation is encountered both in connection with new construction and with repair-replacement decisions on existing facilities. In all cases, the basic question is, "How much are the future earnings (or savings) worth?" Previous sections of this chapter discussed the various factors that affect a project's IRR. The two most important factors in determining the IRR of a prospective investment are: (1) the operating profits generated by the investment, and (2) the capital investment required.

Although economic life has less of an impact on IRR than either of these factors, it is still an important consideration in choosing between alternative investment opportunities.

IMPORTANCE OF ECONOMIC LIFE

Selecting an appropriate economic life for a proposed project is usually one of the most difficult, and frequently one of the most questionable, aspects of the feasibility study. For municipal or public works projects, the study period may be defined as the period over which the project is to be financed; the project's economic life, then, is assumed to be equal to the life of the bonds.

For industrial projects, though, the study period is probably much shorter than the life of the equipment and facilities under consideration. An industrial firm seldom goes beyond a 10-year operating period in analyzing a new venture, and 5- to 8-year study periods are common. Actually, what happens beyond 10 years makes very little difference in most industrial feasibility studies; but whether the study period is 5 years or 10 years can have a very substantial impact on the apparent IRR.

MAXIMUM IRR FROM CHOICE OF ECONOMIC LIFE

Consider, for example, a company in a 50-percent tax bracket contemplating an investment of $1 million in a project expected to earn $400,000 annually before taxes. If it is assumed that the investment is such that it can be written off (depreciated) over the project's economic life, the depreciation charge is deducted from the $400,000 to obtain the net taxable income, and cash flow is calculated as the net profit after taxes plus the depreciation charge. Over a 5-year period, then, depreciation will be $200,000 per year and the net cash flow will be:

$$(400,000 - 200,000) \ (0.50) + 200,000 = \$300,000$$

A $300,000 net cash inflow for 5 years, resulting from a $1 million initial investment, is equivalent to a 15.2 percent IRR.

Projecting a 10-year economic life would result in a net annual cash inflow of

$$(400,000 - 100,000) \ (0.50) + 100,000 = \$250,000.$$

This cash inflow over a 10-year period is equivalent to a 21.4-percent return.

Extending the project's economic life to 15 years brings about relatively little additional change in the IRR. Over a 15-year period, an average net annual cash flow of $233,000 will be realized, resulting in a yield of 22.1 percent on the invested capital. This is the same yield that would be obtained over a 20-year period.

The top curve in Figure 4.6 illustrates the effect of economic life on the expected rate of return in the situation just described. An unusual feature of the realationship is that as the estimated life increases beyond a certain point—about 17 years in the example—the apparent IRR actually declines. Thus, a 17-year life brings a 22.2-percent return; but with an infinite life,

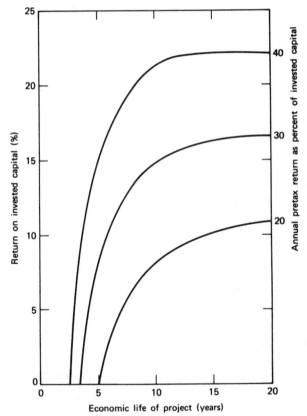

Figure 4.6. *Profitability versus economic life for depreciable investments (50-percent tax bracket).*

annual depreciation will be zero, and net cash flow will be a steady $200,-000, corresponding to an even 20-percent IRR.

There is, in every instance in which income taxes are involved, some intermediate time period at which the return on investment is a maximum; this point might be termed the "optimum economic life."

The two lower curves in Figure 4.6—representing gross annual returns of 30 and 20 percent of the invested capital—show the same general relationships as the previous example, although the IRR is affected more gradually by the project's economic life at these lower levels.

With the $200,000 annual return on a $1 million investment (the bottom curve), the project will just break even at 5 years. At a 10-year life, it will show a return of 8.1 percent; at 15 years, 10.2 percent; and at 20 years, 10.9 percent. At 20 years it will not yet have reached its optimum economic life, but on extending the life indefinitely, the IRR will gradually decline to 10.0 percent.

The same general relationships hold in the middle curve of Figure 4.6, which represents a 30-percent gross annual return on the invested capital. Here the IRR will eventually level off at 15.0 percent, but will reach a maximum of about 17 percent somewhere in the 10–20-year area.

These three examples are all based on a 50-percent tax rate, and an investment that can be fully depreciated over the project's economic life, with a uniform annual series of cash returns, and with depreciation computed on a straight-line basis. Employing accelerated depreciation methods would increase the apparent IRR by a percent or two, but would not alter the general relationships.

TAX-FREE INVESTMENTS

For a tax-free investor, depreciation has no effect on cash flow, and the net cash flow is the same as the gross annual return. In this case, the IRR increases as the project life increases. A 10-percent per year net return on the invested capital will break even in 10 years; it will earn 5.6 percent over 15 years, 7.8 percent over 20 years, 9.3 percent over 30 years, 9.8 percent over 40 years, and approach 10.0 percent as the time period is lengthened infinitely. In this case, there is no "optimum" economic life in terms of the return on investment. Figure 4.7 illustrates the relationship between IRR and net annual returns in a tax-free situation.

The main distinctions between feasibility studies for industrial versus public clients are:

1. Industrial projects consider income taxes, to which public projects are not subjected.
2. Because of the income tax structure, a rapid depreciation method is important in industrial studies, whereas longer-term depreciation may be preferred in studies by public utilities, if their rate calculations vary with book value.
3. Most industrial firms require a return on invested capital above 15 percent; most public works projects are concerned primarily with covering their cost of capital—usually less than 10 percent.
4. For industrial projects, a relatively short payout period is usually necessary, whereas for public projects, it is advantageous to extend the project life over as long a period as possible.
5. Because of all these factors, industrial projects have some identifiable optimum economic life; extending the study period beyond this point reduces the apparent IRR.

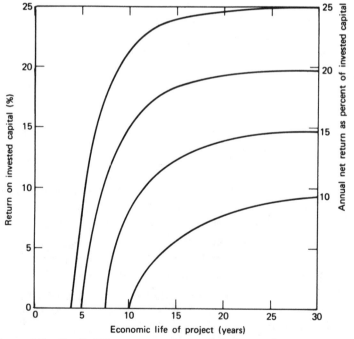

Figure 4.7.   *Profitability versus economic life for tax-free investments.*

INCOME TAXES

Income tax rates, like interest rates, are an important tool for the government in its attempt to control inflation. A high income tax rate decreases the supply of money available for business investment and consumer spending, and high interest rates discourage borrowing (or debt financing) and make new investment appear less desirable from an economic standpoint.

Consequently, income taxes are an important part of most feasibility studies for industrial clients, whereas interest rates may be a controlling factor for public works and other nontaxpaying entities. Privately owned utilities are seriously affected by both tax rates and interest rates, because they usually rely heavily on long-term borrowing to finance new projects.

In the *which* type of feasibility study—that is, studies whose objective is to identify *which* of several alternatives is most economically attractive—income taxes have little effect on the outcome, because the same tax rate applies to all alternatives. In this situation, the most attractive alternative before taxes is likely to be the most attractive after taxes.

But in the *whether* type of feasibility study—where the question is not which way to do it but *whether* to do it at all—income taxes can have a major impact on the investment decision. For example, if a company in a 50-percent bracket requires that new projects yield a minimum return on investment of 15 percent, the pretax return will have to be in the 30-percent range for the investment to quality as a satisfactory commitment.

If tax-exempt municipal bonds were currently yielding 5.5 percent with virtually no risk, a commercial bank's reluctance to loan money even to its prime customers at anything less than 11.0 percent would certainly be understandable.

High income tax rates, then, discourage business investment, and low tax rates encourage investment—essentially the same effect as high or low interest rates.

INCOME TAXES AND INVESTMENT ECONOMICS

In comparing alternative investment opportunities, different cash flows between the alternatives are changed by the amount of the actual tax payable on the differences in accounting profits. Regardless of whether the taxes are assessed on ordinary income or capital gains, they can be considered simply another cash expense or outflow as long as they are calculated according to appropriate legal and accounting standards.

The effect of income taxes on the apparent attractiveness of a proposed investment can be visualized from the simple example shown in Table 4.7. This example shows an initial $1000 investment, recovered through a depreciation charge of $200 annually for 5 years, generating additional revenues of $400 per year over that period. The net profit before taxes is $200 each year, and the corresponding gross cash flow is $400. Net cash flow, of course, decreases as the tax rate increases, ranging from $400 at a zero tax rate to $300 at a 50-percent rate.

INTERNAL RATE OF RETURN

The IRR associated with these tax rates has a maximum of about 29 percent at the zero rate and a minimum of zero at the 100-percent tax rate. The IRR resulting from intermediate tax rates can be interpolated roughly as 15 percent at a 50-percent rate, 18 percent at a 40-percent rate, and 20 percent at a 30-percent rate. In this example, each 10-percent increment in tax rate changes the IRR by about 3 percent.

OTHER EFFECTS OF INCOME TAXES

Income taxes may play an important role in setting financial policy for a company. Since interest payments are deductible from gross income, debt financing has been preferred to equity financing by many established companies, unless equity capital can be obtained by giving up relatively little in present earnings and ownership.

Even personal income taxes can affect, indirectly, a company's cash position, although personal taxes usually do not influence investment policies concerning new projects. In attracting new equity capital, outside investors normally prefer capital gains to dividends because of the difference in tax liabilities; consequently, if the company can reinvest its surplus cash to generate earnings growth and resulting capital appreciation, investors will usually respond by placing a higher value on the company's equities. Thus, the cost of equity capital may be lowered substantially and, with a lower cost of capital, more projects may be considered economically feasible.

Regardless of the effects of income taxes on a company's operations, it is essential that they be given proper consideration in the feasibility study;

Table 4.7   Effect of Income Tax Rate on IRR

Year	Gross Cash Flow	Depreciation	Net Profit before Income Tax	Net Cash Flow After Income Taxes at Various Tax Rates		
				0%	30%	50%
0	(1000)	0	0	(1000)	(1000)	(1000)
1	400	200	200	400[a]	340[a]	300[a]
2	400	200	200	400	340	300
3	400	200	200	400	340	300
4	400	200	200	400	340	300
5	400	200	200	400	340	300
IRR				28.7%	20.7%	15.3%

[a] Depreciation and net profit before income tax (1-tax rate).

this requires a knowledge of income tax laws and regulations, and a familiarity with the company's tax status.

SUMMARY

The DCF technique offers a logical and generally accepted means of relating the cash earning potential of proposed project to its investment requirements. The project's IRR thus measured defines the maximum interest rate at which money can be borrowed to finance the project and subsequently repaid out of cash earnings over the project's economic life. Simple DCF problems may be solved mathematically, although most are far too complex for a mathematical solution. A trial-and-error approach is generally used, usually in conjunction with either algebraic or graphical interpolation. Some of the difficulties associated with DCF analysis can be circumvented by employing graphical techniques. For projects assumed to have uniform cash flows, a simplified approach can be taken; this approach, incorporating just six factors, is especially useful in screening new projects before all the details are fully known. By varying the six factors systematically, the sensitivity of the project's IRR to each factor can be easily estimated. By assigning probabilities to each of the factors, the expected IRR for the project can be calculated, along with the overall range of possible outcomes. Of the six major factors, the capital investment requirement and the gross operating profits generally have the greatest effect on the IRR of a given project. Economic life, however, may be an important factor in comparing different projects to determine which is most desirable; and income taxes exert a major influence on the decision to consider a specific project.

5

Depreciation and
Economic Life

Age will not be defied.

Francis Bacon

In a physical sense, the term "depreciation" refers to the impaired service-ableness of a capital asset because of its age. In an economic sense, it implies a decrease or difference in value from its original cost, also as a result of age. However, in the most realistic sense, depreciation is simply a legal or accounting device defined by the Internal Revenue Service as a means of recovering the original cost of a capital asset through a series of annual recovery deductions. The chief importance of depreciation in economic analysis, therefore, is its effect on income taxes and its subsequent impact on a project's cash flow.

The importance of economic life in a project's cash flows and the effect of cash flows on profitability are covered in Chapter 4. This chapter discusses methods of estimating economic life and computing recovery deductions, and the effect of these deductions on profitability over a project's economic life.

As in the material covered in earlier chapters, the individual's point of view makes a big difference in how depreciation and economic life are perceived. The accountant is most interested in the depreciation of assets already in service, from the standpoint of tax-deductible expenses. The engineer, on the other hand, is probably more interested in the depreciation of assets not yet acquired for use in new ventures, with an eye toward capital recovery. Both the accountant and the engineer need to recognize the role of depreciation in generating cash flows, and the impact on profitability of whatever options are legally available.

THE ACCELERATED COST RECOVERY SYSTEM

The Economic Recovery Act of 1981(ERTA) incorporated provisions for the Accelerated Cost Recovery System (ACRS), which represented a major change in tax accounting for capital expenditures and offered substantial incentives for businesses to make new capital investments. The ACRS replaced the useful-life depreciation rules that had been developed over the preceding half-century, and accelerated the recovery of capital expenditures in two ways:

1. By elimating the useful-life concept and replacing it with a shorter (and audit-proof) recovery period.
2. By allowing larger recovery deductions during the early years of an asset's depreciable life.

Under these later tax laws, businesses take "recovery deductions" (instead of depreciation) on capital expenditures qualifying as "recovery property." Recovery property is defined as any depreciable, tangible property used in a business or held for the production of income. Only such property placed in service during or after 1981 qualified for accelerated cost recovery.

It is no longer necessary for tax purposes to determine the useful life and salvage value of assets or to elect the "Class Life Asset Depreciation Range System." Instead, the ACRS establishes several property classes and specifically defines the annual deductions to be allowed as percentages of the original cost of the property.

Under ACRS, qualifying capital expenditures are generally recovered over a period of 3, 5, 10, or 15 years.

THREE-YEAR PROPERTY CLASS.   The three-year property class includes automobiles, light trucks, and certain personal and other property that formerly had an Asset Depreciation Range (ADR) class life of 4 years or less. It also includes personal property used in research and development, along with some special tools.

FIVE-YEAR PROPERTY CLASS.   The five-year property class covers most machinery and equipment, and also includes whatever personal property is not classified elsewhere.

TEN-YEAR PROPERTY CLASS.   The ten-year property class includes certain public utility property and short-lived real property having an ADR class life of less than 13 years. Such items as amusement park property, railroad tank cars, and manufactured homes fall into this category.

FIFTEEN-YEAR PROPERTY CLASS.   The fifteen-year class consists largely of longer-lived, public utility, tangible personal property and of certain real property.

QUICK RECOVERY

The allowable recovery deductions for a particular asset depend both on the recovery method and the recovery period, with a single recovery method and period applying to all property within a class placed in service in the same taxable year. Certain exceptions apply to real estate, which is handled on a property-to-property basis.

The recovery deductions are usually accelerated according to the IRS-published accelerated recovery tables, which provide annual recovery percentages based on the original cost of the property. The recovery percentage for the first year (the year of acquisition) allows only a half-year's deduction, and there is no recovery deduction for the year of disposition.

Three separate accelerated recovery schedules apply, depending on the year in which the assets were placed in service. The first (Table 5.1) applies to property placed in service during the 1981–1984 period; the second (Table 5.2) is for assets placed in service in 1985; and the third (Table 5.3) covers property installed from 1986 until whenever the regulations are changed again.

This point, incidentally, is an extremely important one. Tax laws and regulations do change occasionally, because of prevailing economic or po-

Table 5.1 Accelerated Recovery Deductions for Property Placed in Service from 1981 to 1985

Recovery Year	Property Class (%)			
	3-Year	5-Year	10-Year	15-Year
1	25	15	8	5
2	38	22	14	10
3	37	21	12	9
4		21	10	8
5		21	10	7
6			10	7
7			9	6
8			9	6
9			9	6
10			9	6
11–15				6

Table 5.2 Accelerated Recovery Deductions for Property Placed in Service During 1985

Recovery Year	Property Class (%)			
	3-Year	5-Year	10-Year	15-Year
1	29	18	9	6
2	47	33	19	12
3	24	25	16	12
4		16	14	11
5		8	12	10
6			10	9
7			8	8
8			6	7
9			4	6
10			2	5
11				4
12				4
13				3
14				2
15				1

Table 5.3   Accelerated Recovery Deductions for Property Placed in Service During 1986 and Later

Recovery Year	Property Class (%)			
	3-Year	5-Year	10-Year	15-Year
1	33	20	10	7
2	45	32	18	12
3	22	24	16	12
4		16	14	11
5		8	12	10
6			10	9
7			8	8
8			6	7
9			4	6
10			2	5
11				4
12				3
13				3
14				2
15				1

litical pressures, and "ignorance of the law is no excuse." Anyone engaged in economic analysis must keep currently informed regarding whatever regulations might affect the projects being analyzed.

Utilizing the 1981–1984 accelerated recovery rates is roughly comparable to employing the 150 percent declining balance method up to the optimum crossover point, then switching to the straight-line approach. The 1985 rates follow a 175 percent declining balance, with a switch to the sum-of-the-years-digits method. Recovery deductions for 1986 and later approximate a combination of a 200 percent declining balance for about the first half of the asset life, with a subsequent change to the sum-of-the-years-digits technique.

The taxpayer may choose to take longer recovery periods than those defined in the accelerated recovery tables, with recovery over the optional periods based on the straight-line method. Briefly stated, 3-year property can optionally be recovered over 3, 5, or 12 years on a straight-line basis; 5-year property over 5, 12, or 25 years; 10-year property, over 10, 25, or 35 years; and 15-year property, over 15, 35, or 45 years.

Investment tax credits may play an important role in studies involving new ventures. In general, tax credits of 6 percent are allowed on 3-year recovery class property, and 10 percent on qualifying property in other classes.

A number of other rules and regulations may apply to a particular project, depending on the type of project and the kinds of assets involved. There are specific provisions covering component depreciation, antichurning rules, partial expensing of certain recovery property, tax preference items, minimum taxes, foreign recovery property, leasehold improvements cost recovery, "at-risk" rules, research and experimentation credit, and many other items that could be important in a specific venture.

ESTIMATING ECONOMIC LIFE

Even though an asset's physical or economic life has little bearing on its depreciable life for income tax purposes, the life span is still an important consideration in setting the time period over which a proposed venture is to be examined.

Obviously, the accuracy of such a study is largely dependent upon the choice of an appropriate economic life for the asset or project. This estimate of economic life, along with whatever method is used to compute depreciation, provides the basis for a major portion of the asset's annual ownership cost. The time period over which the study is to be made exerts a significant influence on a proposed investment's apparent profitability, feasibility, or desirability.

The entire amount of an investment can be recaptured through the annual recovery deductions. However, it is the timing of these deductions—reflecting the time value of money—that largely determines whether or not the investment is a good one. Although caution dictates that a conservative estimate of economic life be used, especially in studies in which a relatively high degree of risk may be encountered, the ultimate objective is to have the estimate fit the experience of the particular property in question. In virtually every case, the depreciable life will be shorter than the economic life.

There are several different ways to estimate the economic life of capital assets. The most sophisticated methods involve actuarial analysis. Less sophisticated, but extremely useful and far more convenient, are the old IRS guidelines that were used as a primary basis for computing depreciation until the accelerated cost recovery provisions of the Economic Recovery Tax Act of 1981 went into effect.

IRS GUIDELINES.   Internal Revenue Service Publication No. 456, *Depreciation Guidelines and Rules,* provides tables of the useful lives of a variety of asset types, arranged by industry groups. Tables 5.4 through 5.7 summarize the guidelines developed by the IRS for various types of capital assets used in business and industry.

Table 5.4   Useful-Life Guidelines for Assets Used by Business in General

Type of Asset	Economic Life (Years)
Office furniture, fixtures, machines, and equipment	10
Transportation equipment	
Aircraft	6
Automobiles	3
Buses	9
Light general-purpose trucks	4
Heavy general-purpose trucks	6
Railroad cars	15
Over-the-road tractor units	4
Trailers and trailer-mounted containers	6
Water transportation equipment	18
Land improvements	20
Buildings	
Apartments	40
Banks	50
Dwellings	45
Factories	45
Garages	45
Grain elevators	60
Hotels	40
Loft buildings	50
Machine shops	45
Office buildings	45
Stores	50
Theaters	40
Warehouses	60

Table 5.5   Useful-Life Guidelines for Assets Used in Nonmanufacturing Activities

Type of Asset	Economic Life (Years)
Agriculture	
Machinery and equipment	10
Animals	
Cattle, breeding or dairy	7
Horses, breeding or work	10
Hogs, breeding	3
Sheep and goats, breeding	5
Farm buildings	25
Contract construction	
General contract construction	5
Marine contract construction	12
Logging and sawmilling	
Logging	6
Sawmills	10
Portable sawmills	6
Mining	10
Recreation and amusement	10
Personal and professional services	10
Wholesale and retail trade	10

Use of the IRS guidelines was never mandatory; they were intended solely as a guide to "what might be considered reasonably normal periods of useful life." In practice, the service lives used as the basis for depreciation accounting were determined by the asset's particular operating conditions and experience, tempered by judgment concerning the probability of technological improvements and economic changes. Although the tax laws may change, the necessity for sound judgment remains.

ACTUARIAL ANALYSIS

Actuarial approaches to estimating service life involve the construction of frequency curves, survivor curves, and probable life curves. These curves for physical property are similar in both derivation and use to the mortality

Table 5.6 Useful-Life Guidelines for Assets Used in Manufacturing

Type of Asset	Economic Life (Years)
Aerospace industry	8
Apparel and fabricated textile products	9
Cement manufacture	20
Chemicals and allied products	11
Electrical equipment	
Electrical equipment	12
Electronic equipment	8
Fabricated metal products	12
Food and kindred products	12
Glass and glass products	14
Grain and grain mill products	17
Knitwear and knit products	9
Leather and leather products	11
Lumber, wood products, and furniture	10
Machinery, except electrical and metalworking	12
Metalworking machinery	12
Motor vehicles and parts	12
Paper and allied products	
Pulp and paper	16
Paper finishing and converting	12
Petroleum and natural gas	
Drilling, geophysical, and field services	6
Exploration, drilling, and production	14
Petroleum refining	16
Marketing	16
Plastics products	11
Primary metals	
Ferrous metals	18
Nonferrous metals	14
Printing and publishing	11
Professional, scientific, and other instruments	12
Railroad transporation equipment	12
Rubber products	14
Ship and boat building	12
Stone and clay products, except cement	15

Table 5.6   (*Continued*)

Type of Asset	Economic Life (Years)
Sugar and sugar products	18
Textile mill products, except knitwear	
Textile mill products	14
Finishing and dyeing	12
Tobacco and tobacco products	15
Vegetable oil products	18
Other manufacturing	12

curves prepared by insurance companies to reflect human life expectancy. The survivor curve for physical property is developed from historical data on similar property, just as the mortality curve is developed by actuarial analysis of vital statistics on humans.

By studying the retirement history of a particular type of property, a frequency distribution of property retirement can be obtained. From this, a cumulative frequency distribution curve—or survivor curve—can also be plotted, as can a probable life curve indicating the potential future life of an asset of any given age. Figure 5.1 shows a generalized survivor curve, with the corresponding frequency distribution and probable life curves. The frequency curve shows the percentage of the total number of units retired during each time interval. The survivor curve shows the percentage of the total number of units of a given age still in service. The probable life curve gives the total life expectancy of an asset of any given age. The horizontal distance between survivor and probable life curves represents the remaining life of an asset of any age.

Survivor curves are most useful when dealing with "mass" properties, where large numbers of items of relatively low unit value are involved. Typical examples of mass properties include gas, electric, and water meters, distribution mains, utility poles, and similar items.

Just as in analyzing human life expectancy for insurance purposes, analyzing mass properties for depreciation purposes can have value in planning.

The probable future life of an asset can be statistically predicted from its present age, thus providing a useful guideline for deciding at what age it should be retired or replaced—as with humans, whenever the present value of its future services is less than its replacement cost.

Table 5.7   Useful-Life Guidelines for Assets Used in Transportation, Communications, and Public Utilities

Type of Asset	Economic Life (Years)
Air transport	6
Central steam production and distribution	28
Electric utilities	
Hydroelectric plant	50
Nuclear production plant	20
Steam production plant	28
Transmission and distribution facilities	30
Gas utilities	
Distribution facilities	35
Manufactured gas production plant	30
Natural gas production plant	14
Trunk pipelines and related storage facilities	22
Motor transport—freight	8
Motor transport—passengers	8
Pipeline transportation	22
Radio and television broadcasting	6
Railroads	
Machinery and equipment	14
Structures and improvements	30
Wharves and docks	20
Power plant and equipment	
Hydraulic generating equipment	50
Nuclear generating equipment	20
Steam generating equipment	28
Steam and other power plant equipment	28
Water transportation	20
Water utilities	50

DEPRECIATION, ECONOMIC LIFE, AND THE IRR

Regardless of the schedule used for taking the allowable recovery deductions, the entire amount of the qualifying investment in capital assets will be recovered; only the timing of the capital recovery will be affected. Any

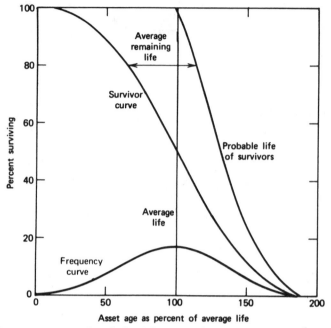

Figure 5.1.   *Survivor and probable life curves for mass properties show their life expectancies.*

change in timing will, in turn, change the annual cash flow patterns and will subsequently alter the project's IRR.

Table 5.8 shows a variety of recovery deduction schedules that might be possible for a $100,000 investment in a new venture having an estimated economic life of 10 years. The different depreciation schedules include accelerated cost recovery over 3, 5, and 10 years, and straight-line depreciation over 5 and 10 years.

Since the shorter term recovery methods allow greater write-offs during the project's early years, they reduce income tax liabilities and thereby enhance the early years' cash flows. Table 5.9 shows the net cash flows associated with each of these capital recovery options, assuming the project earns a $50,000 net income before recovery deductions and income taxes each year of its life. In each case, the total pretax, predepreciation earnings over the entire 10-year period would be $500,000. Deducting the $100,000 initial qualifying investment from this amount leaves $400,000 in taxable income, or $200,000 in net profits, after taxes are deducted at a 50 percent rate. Adding back the recovery deductions gives the project a net cash flow of $300,000 over the 10-year study period.

Table 5.8   Annual Recovery Deductions for a $100,000 Qualifying Investment

Year	Accelerated Recovery			Straight-Line Recovery	
	3-Year	5-Year	10-Year	5-Year	10-Year
1	33,000	20,000	10,000	20,000	10,000
2	45,000	32,000	18,000	20,000	10,000
3	22,000	24,000	16,000	20,000	10,000
4	0	16,000	14,000	20,000	10,000
5	0	8,000	12,000	20,000	10,000
6	0	0	10,000	0	10,000
7	0	0	8,000	0	10,000
8	0	0	6,000	0	10,000
9	0	0	4,000	0	10,000
10	0	0	2,000	0	10,000
Total	100,000	100,000	100,000	100,000	100,000

Table 5.9   Net Cash Flows for a $100,000 Investment[a]

Year	Accelerated Recovery			Straight-Line Recovery	
	3-Year	5-Year	10-Year	5-Year	10-Year
0	−100,000	−100,000	−100,000	−100,000	−100,000
1	41,500	35,000	30,000	35,000	30,000
2	47,500	41,000	34,000	35,000	30,000
3	36,000	37,000	33,000	35,000	30,000
4	25,000	33,000	32,000	35,000	30,000
5	25,000	29,000	31,000	35,000	30,000
6	25,000	25,000	30,000	25,000	30,000
7	25,000	25,000	29,000	25,000	30,000
8	25,000	25,000	28,000	25,000	30,000
9	25,000	25,000	27,000	25,000	30,000
10	25,000	25,000	26,000	25,000	30,000
Total	200,000	200,000	200,000	200,000	200,000
IRR	33.5%	31.5%	28.7%	30.6%	27.3%

[a] Results are from a 10-year study period.

The IRRs associated with each of the five options described in the Tables 5.8 and 5.9 range from 33.5 percent for the 3-year capital recovery period to 27.3 percent for the 10-year straight-line option. The 5-year accelerated recovery schedule yields an IRR of 31.5 percent; 10-year accelerated recovery returns 28.7 percent annually; and 5-year straight line depreciation results in a 30.6 percent rate of return under the prescribed conditions.

The conclusion is obvious enough: the shorter the capital recovery period, the higher the return on investment. If, in the preceding example, the cut-off point for new venture approval were set at an IRR of 30 percent, the depreciation schedule alone could determine whether or not the project would be pursued.

Although the capital recovery schedule is important, the project's economic life (or study period) has an even greater impact on its IRR.

Table 5.10 shows the same situation, except that the study period has been reduced from 10 to 5 years. Three of the same capital recovery options are presented here: 3- and 5-year accelerated recovery, and 5-year straight-line depreciation. In each case, the total of net cash flows over the 5-year study period is the same: 5 years of pretax, predepreciation income of $50,-000, less $100,000 in recovery deductions, with the balance taxed at 50 percent, leaving $75,000 in net profits after taxes. Adding this figure back to the recovered capital gives a 5-year total net cash flow of $175,000.

Again, the timing of these cash flows results in a different IRR for each case. Their rankings are the same as before, but the yields are considerably lower than for the 10-year study period. The 3-year accelerated recovery

Table 5.10   Net Cash Flows for a $100,000 Investment[a]

	Accelerated Recovery		Straight-Line Recovery
Year	3-Year	5-Year	5-Year
0	−100,000	−100,000	−100,000
1	41,500	35,000	35,000
2	47,500	41,000	35,000
3	36,000	37,000	35,000
4	25,000	33,000	35,000
5	25,000	29,000	35,000
Total	75,000	75,000	75,000
IRR	25.3%	23.1%	22.1%

[a] Results are from a 5-year study period.

schedule returns 25.3 percent annually on the invested capital; 5-year ac-celerated recovery deductions generate an IRR of 23.1 percent; and, with 5-year straight-line depreciation, the project earns 22.1 percent.

A comparison of the IRR generated by these different examples, employ-ing different recovery deduction schedules and study periods, looks like this:

Recovery Schedule	5-Year Study Period (% IRR)	10-Year Study Period (% IRR)
3-Year accelerated	25.3	33.5
5-Year accelerated	23.1	31.5
5-Year straight line	22.1	30.6
10-Year accelerated	—	28.7
10-Year straight line	—	27.3

Obviously, both the capital recovery schedule and the economic life of a project are important determinants of its profitability, although the length of the study period has a far greater impact than the depreciation period. A new venture, especially, should not be penalized by assuming its economic life to be the same as its depreciable life. In general, recovery deductions should be taken as quickly as possible, and the project's economic life should be estimated as accurately as possible.

SUMMARY

Depreciation plays an important role in economic evaluations; it provides the means by which the out-of-pocket cost of a qualifying capital asset can be recovered through a series of annual recovery deductions, usually within a time period considerably shorter than the asset's economic life. The eco-nomic life of an existing asset, or of assets similar to those for which experi-ence data have been accumulated, can be calculated by employing actuarial techniques. In new ventures, either past experience, or the IRS guidelines formerly used as a basis for depreciation, can provide a reason-ably sound basis for estimates of an asset's economic life. Both depreciation and economic life have significant impacts on a project's apparent profit-ability, as measured by its IRR, although the effect of the venture's economic life is considerably greater than the effect of its capital recovery schedule. Maximum profitability results from taking the fastest allowable recovery deductions, and analyzing the project over a realistically appropriate time period.

6

The Cost
of Capital

The fact is, in my opinion,
that we often buy money
very much too dear.

W. M. Thackeray

The cost of capital is an essential part of nearly every economic evaluation.
It is particularly important in providing a valid test of project desirability
and in helping to select the best available use for a firm's money. The IRR
must obviously be at least equal to the cost of capital used to finance the
new venture.

Capital can be defined as the funds used in the operation of a business.
Most firms have two sources of capital: debt and equity capital. Debt capital is obtained by borrowing, whereas equity capital is supplied by the
firm's owners, or stockholders, and represents the amount of their ownership in the company. Equity capital also includes retained earnings. Those
who supply the firm's capital—whether lenders or owners—naturally ex-

pect a return on their money. Lenders receive their return in the form of interest. Shareholders expect to receive theirs as a combination of dividends and capital appreciation.

Thus, the cost of capital has different meanings to different people. To the lender, it signifies the interest he receives on the money he lends. To the stockholder, it represents his income from dividends and capital appreciation. To the company, the cost of capital is the amount that must be earned to satisfy the expectations of the lenders and shareholders.

The company can determine its cost of capital by measuring the cost of each type of capital employed in the business and weighing those costs according to their relative contribution. The resultant figure represents the company's overall cost of money. This overall cost of capital is an appropriate figure to use as a basis for comparisons among projects in most economic feasibility studies. Should outside sources of capital be used for financing a specific venture, then the amount paid for only that capital can be used in assessing the proposed venture's merits. Even so, new projects that are expected to earn less than the company's overall cost of capital are, at best, of questionable merit and are, more likely, disastrous.

The financial characteristics and point of view of the investing firm determine how the cost of capital is used in a specific study. A relatively small firm with a single project under consideration is primarily concerned with the immediate sources of capital for that particular project. If the project is expected to yield a return significantly in excess of the cost of borrowed money, the investment is apt to be viewed favorably.

The large corporation is faced with a more complex problem. Often, the financing of a particular venture from specific sources has a secondary effect on the capability of the corporation to raise capital for other projects. For this reason, it is usually better to consider the problems of raising capital, and the problems of determining the expected return of a specific venture, as completely independent and distinct.

SOURCES OF FUNDS

The cost of capital used by a company in financing a new venture depends largely upon how the project is financed and the source of the funds utilized. It is therefore important to consider all the financial details of a project before passing judgment on its desirability or feasibility.

In obtaining the necessary funds for a major project, the company must carefully weigh the outside sources available to it. The characteristics of the various segments of the capital and money markets to which a corporation, or in some cases a government body, can turn to obtain its funds are discussed in Chapter 2. The major financial markets available to corporations include corporate bonds, corporate equities, commercial bank loans, and commercial paper.

Corporate bonds, equities, and bank loans constitute the most important sources of long-term capital used in financing major new ventures. Bank loans and commercial paper are employed extensively as sources of relatively short-term working capital.

THE COST OF CAPITAL FROM BOND SALES INVOLVES THE TIME VALUE OF MONEY

The cost of short-term borrowings on the money market can be determined easily by reference to the current rates on commercial loans, treasury bills, and other short-term obligations. Bonds present a more difficult problem. Investors buy long-term bonds primarily for their annual interest returns, whereas short-term bonds may be purchased either for their capital appreciation potential or as a depository for idle funds at a reasonable interest rate until a better opportunity comes along. A bond's overall value as an investment, or its cost as a source of funds, includes both of these elements measured in terms of a composite annual percentage interest over the remaining life of the bond.

The annual interest cost of a bond is determined by four factors:

1. Current market price.
2. Redemption or maturity value.
3. Coupon rate or annual interest payment.
4. Years to maturity.

Market price is the only variable among these four factors, since the annual interest payments are defined by a bond's coupon rate, and the maturity date and value are specified. Consequently, the bond's price is the only factor that can be varied in the marketplace to reflect changing interest rates and money costs.

Given price, coupon rate, and maturity date, two approaches can be taken in establishing the attractiveness of a bond as an investment: (1) present value and (2) DCF.

Present value of a bond

The would-be buyer of a bond could consider its present value, which consists of the present value of its maturity value plus that of its annual interest payments. The equation for a bond's present value is:

$$P = \frac{M}{(1 + R)^N} + C \left[\frac{(1 + R)^N - 1}{R(1 + R)^N} \right]$$

where P is the present value; M is the redemption or maturity value; C is the annual interest payment determined by the coupon rate; R is an appropriate discount rate, reflecting the short- and long-term money market rates; and N is the number of years until the bond matures.

Using this equation, a \$1000 bond maturing in 16 years and bearing a 5-percent coupon (yielding \$50 annually) is, at a 7-percent discount rate, worth

$$P = \frac{1000}{(1.07)^{16}} + 50 \frac{(1.07)^{16} - 1}{0.07(1.07)^{16}}$$

$$= 339 + 472$$

$$= \$811$$

The present value of this bond's redemption value is \$339, and the annual interest payments are worth \$472 at the 7-percent discount rate. Under these conditions, if the bond is sold for less than \$811, it will yield (or cost its issuer) more than 7 percent annually. Similarly, if the bond is priced above \$811, it will yield or cost less than 7 percent annually over its remaining life.

Discounted cash flow

The seller of bonds is concerned with a composite of interest and face values that can best be computed by employing the DCF technique. This approach does away with the necessity of artificially establishing a discount rate and provides a single index for comparing bonds of similar quality but having different prices, coupon rates, and maturity dates. This method is used in developing the values published in the bond yield tables utilized by bond brokers and investors.

The DCF technique is a special case of the present value approach described in the preceding section. Instead of selecting an arbitrary discount rate and computing the bond's present value, the DCF method selects the bond's present value and computes the discount rate that satisfies the prescribed conditions. The bond's present value is assumed to be its current market price, and the computed discount rate becomes, by definition, the annual percentage yield or annual interest cost. Thus, the equation is the same, only the unknown is different.

Using the example of a $1000 bond maturing in 16 years, carrying a 5-percent coupon, and priced at $740, the equation becomes

$$\$740 = \frac{1000}{(1+R)^{16}} + 50 \left[\frac{(1=R)^{16} - 1}{R(1+R)^{16}} \right]$$

where R represents the annual percentage return when the bond is held to maturity. Solving for R in an equation such as this is a laborious task, generally accomplished by trial and error. In this case, R is about 7.9 percent:

$$P = \frac{1000}{(1.079)^{16}} + 50 \left[\frac{(1.079)^{16} - 1}{0.079(1.079)^{16}} \right]$$

$$= 296 + 445$$

$$= \$741$$

which is close enough to the $740 market price. The cost of the issuing corporation's long-term debt capital here, then, is 7.9 percent.

If one lacks the patience, or a computer, to go through the calculations associated with the DCF approach, the yield to maturity (YTM) of a bond can be estimated by using the approximation:

$$\text{YTM} = \frac{2C}{P+M} + \left(\frac{M}{P} \right)^{1/N} - 1.00$$

where, as before, C is the annual coupon rate or interest payment per $1000 maturity value, M is the maturity value, N is the number of years until maturity; and P is the current market price. Using the same example, the approximate YTM is:

$$\text{YTM} = \frac{2 \times 50}{740 + 1000} + \left(\frac{1000}{740} \right)^{1/16} - 1.00$$

$$= 0.057 + 1.019 - 1.00$$

$$= 0.076, \text{ or } 7.6 \text{ percent}$$

which compares favorably with the actual 7.9 percent determined previously. If bond yield tables are not available, this approximate method is useful either for obtaining a quick and easy measure of a bond's investment value or interest cost, or for establishing a trial rate to use in subsequent DCF calculations.

THE COST OF EQUITY CAPITAL

The cost of equity capital includes more than dividends. Although there is no contractual obligation on the company's part to pay dividends, raising money by selling stocks is a real economic cost of doing business. Selling stock does, however, create an obligation to earn a given amount. If a publicly held company is to maintain its ability to attract equity capital, the company must earn for its investors at least as much as they could earn by investing their money elsewhere at comparable risk. Should the company cease to be attractive to equity investors, it is likely to become equally unattractive to other sources of capital.

Estimating a realistic cost for equity capital requires considerable judgment. The price of a company's common stock reflects the investing public's appraisal of the present worth of the company's future earnings. The investor is, in effect, purchasing an annuity made up of a stream of future dividends or earnings, and an opportunity to sell his shares whenever he wishes at the prevailing market price.

Thus, the cost of equity capital includes two elements: dividends and the possibility of capital appreciation; both in turn are related to earnings. Just how to combine these two elements into a single figure representing the cost of equity capital has always been a preplexing problem. There are many ways to approach this dilemma, but only two of the simplest are presented here. Both techniques are commonly used.

One approach uses the ratio of present earnings to present market price, on the assumption that the purchase of a share of equity stock at a given price will return to the investor a certain amount per year in earnings. Or, from the company's standpoint, a certain amount of present earnings must be given up to attract each dollar of present equity capital. This ratio must therefore represent the cost of equity capital to the company.

Another means of estimating the cost of equity capital is from the investor's standpoint, rather than the company's. The prospective purchaser of a share of capital stock expects to receive a definite cash return on his investment—the dividend. He probably also expects the stock's value to increase with time, as the company's retained earnings are reinvested in the business to generate more earnings.

Adding the dividend yield to the anticipated earnings growth rate gives a measure of the investor's expectations. For example, if the company pays a 5-percent dividend, and earnings are expected to increase at a rate of 5 percent annually, then 10 percent could be said to represent the amount that the company must earn to satisfy the investor and, therefore, is its cost of capital. This method is probably as justifiable as any other.

A low earnings/price ratio (or high price/earnings ratio) is generally associated with a high anticipated growth rate, and low dividend yields are characteristic of fast-growing companies. Consequently, equity capital may be cheaper for newer, fast-growing firms, while debt capital is usually cheaper for mature companies with high dividend rates.

THE OVERALL COST OF CAPITAL

After a company's capital structure and the cost of each type of capital used in the business have been determined, the overall cost of capital can be calculated. The financial structure of major companies in different fields varies widely, as shown in the following table.

| Industry | Capital Structure (%) | | | |
	Debt Capital	Equity Capital	Other	Total
Oil	17.6	79.0	3.4	100.0
Steel	27.2	67.2	5.6	100.0
Utility	44.2	52.6	3.2	100.0
Chemicals	5.2	82.9	11.9	100.0
Manufacturing	2.7	93.0	4.3	100.0

The cost of capital to these large corporations can be expected to vary equally widely, as do their preferred methods of financing.

Consider, for example, a company having the following financial structure.

	Percent of Total Capital	After-Tax[a] Cost (%)	Weighted Cost (%)
Short-term debt	5.0	6.0	0.30
Long-term debt	15.0	8.0	1.20
Shareholders' equity	80.0	10.0	8.00
Total	100.0		9.50

[a] Interest paid is deductible from gross income; dividends are not.

The company's short- and long-term debt costs can be easily obtained from published financial statements. Shareholders' equity includes all capital stock—common and preferred—and also retained earnings and other long-term reserves.

The cost of conventional debt financing is the average market rate expected on new borrowings and is shown in the example at 6.0 percent for short-term and at 8.0 percent for long-term obligations (both after tax). The exact figures can be established from the current market value of the firm's outstanding notes and bonds and may be well over 10 percent, depending on the prevailing money and capital markets. Equity capital is assumed to be priced at about 10.0 percent in the example.

As shown, weighting each type of capital by its cost results in an overall figure of 9.50 percent.

Should the company's capital structure change, or should the interest rates or other factors relating to the cost of a particular source of capital vary, then the company's overall cost of capital would change accordingly.

If, in the preceding example, the company's financial structure were 10 percent short-term debt, 30 percent long-term debt, and 60 percent equity capital, its overall cost of capital would drop to 9 percent. This company obviously would prefer debt to equity financing.

In the present value approach to measuring profitability, the cost of capital serves as the appropriate discount rate for future earnings; in DCF analysis it might represent the cutoff point beneath which investment proposals would be rejected. Regardless of the method used in appraising new investment opportunities, a realistic measure of investment desirability cannot be obtained without considering the company's overall cost of capital.

SUMMARY

Capital represents the funds used in the operation of a business, and its cost provides a useful guideline in evaluating project desirability and in selecting appropriate uses for available money. A company's cost of capital depends on its capital structure, methods of financing, and sources of funds as well as on the general economic conditions reflected in the capital and money markets. Capital can be obtained either by borrowing (debt capital) or by selling ownership (equity capital). Equity capital is especially important for new or fast-growing companies; debt capital is the less costly and generally preferred source of outside funds for well-established firms. The expected return on new projects should be higher than the cost of capital used in directly financing the project. If internal funds are employed, the proposed project should earn at a rate higher than the company's overall cost of capital. Projects expected to earn at or less than the company's cost of capital should be avoided whenever possible.

7

Cost Analysis, Budgets, and Benefits

Budgets are not merely affairs of arithmetic, but in a thousand ways go to the root of the prosperity of individuals, the relation of classes and the strength of kingdoms.

Gladstone

Every business sells either products or services and in its operation must incur certain costs. The analysis of these costs—including their identification, measurement, allocation, and control—is an extremely important activity. Regardless of where the cost analysis is performed, its purpose is essentially the same: to provide factually accurate and objective information useful to the pursuit of the firm's objectives. The specific purpose of a cost analysis may be to provide information to be used in the design of a

111

rate schedule for a public utility, in the development of an appropriate and reasonable pricing policy for a manufacturer, or as a guide for establishing rental or leasing terms in a real estate venture. It may be utilized as a basis for preparing engineering economic studies, as a planning tool in sales promotion activities, or as an aid in cost control and capital budgeting.

ELEMENTS OF COST

In an accounting sense, cost refers primarily to the expense items on the firm's operating statement such as property taxes, interest, and depreciation, in addition to materials, operating labor, maintenance, supplies, and other conventional expense items. However, the accountant may exclude from the definition of cost the return on equity investment (or profit) as well as the income taxes associated with this return.

The engineering approach to cost also includes an appropriate return or profit on the capital invested in the venture. This approach is necessary in evaluating the economics of a project. Cost, to be completely realistic, must refer to the *entire* economic activity; total cost, then, represents the firm's revenue requirement, the total amount to be received from sales of its goods or services. Cost analysis shows where the money comes from, where it goes, and why.

Only through this engineering concept of cost can the reasonableness of a firm's rates or prices be appraised, as measured by the criterion of whether or not a fair return or profit has resulted after all capital and operating charges are deducted from total revenues.

In serving its customers or clients, a firm performs a variety of different operations, each of which fulfills a specific function and contributes a specific cost. Functionally different operations may be performed at the same physical location, and similar functional operations may be performed at different physical locations. In making a cost analysis, primary emphasis should be on the functional aspects; costs should therefore be grouped by function rather than by physical location.

COSTS MAY BE FIXED, VARIABLE, OR IN-BETWEEN

Cost accounting systems sometimes obscure the relationships between sales, costs, and profits. To help clarify these relationships, costs are usually broken down into either two or three broad categories: fixed, variable, and, sometimes, semivariable.

FIXED COSTS.   Fixed or overhead costs are those whose magnitude depends upon the capacity of a system or upon the time during which an activity is conducted within the system. Charges against a product that arise from fixed costs can be compared to rent. The rent is charged for the time during which a facility and its associated equipment are used to prepare the product for sale. Fixed, or indirect, costs may be prescribed by contract (such as office rent) or incurred in the operation of an organization (such as accounting and personnel departments). Alternatively, fixed charges may consist of a utility's capacity costs associated with investments in its plants or distribution lines.

VARIABLE COSTS.   Expenses that move in close proportion to changes in the level of production—that can be attributed to a specific function and are, therefore, chargeable in their entirety against that particular function—are known as variable or direct costs. An electric generating plant's fuel cost, for example, varies directly with the net generation. Similarly, a gas distribution utility might refer to its "commodity" costs—the cost of natural gas purchased from the supplier's pipeline—as a variable cost. Some firms may have relatively few costs that can be considered completely variable; other businesses may find a large proportion of their total costs to be of a variable nature. In a manufacturing firm, variable costs consist mainly of materials and direct labor.

SEMIVARIABLE COSTS.   In most operations, costs can be categorized as either fixed or variable. Some costs, however, may not fit well into either of these categories, but assume some of the characteristics of each. These must be classified as semivariable. An example is direct or supervisory labor. Semivariable costs are often associated with a firm's need to keep a portion of its physical facilities and personnel intact, almost regardless of the level of business. Some companies must keep a nucleus of competent professional personnel on its payroll at all times to maintain a capability to do business. Unutilized professional staff time thus becomes an overhead or fixed cost, while other staff salaries are variable costs. For example, engineers may be temporarily assigned to a manufacturing facility for the start-up of a new product or to help rescue an ailing one. A utility's customer costs might also be considered semivariable costs, consisting of accounting, collecting, and sales promotion expenses.

THE COST ANALYSIS PROCEDURE

The cost analysis procedure involves four separate and distinct phases:

1. Break down total costs by functional locations.
2. Classify cost elements according to their causes.
3. Determine cost responsibilities.
4. Allocate costs according to the responsibilities involved.

COST BREAKDOWN.   The initial cost breakdown is by function. All costs are presumably incurred for some specific purpose; the functional breakdown of costs simply defines the purpose for which each cost is incurred.

COST CAUSATIONS.   All costs can be attributed to a governing body's capacity to serve its patrons, to the amount of service it actually provides, or to a combination of these two elements. As in manufacturing, all such costs can be classified as fixed, variable, or semivariable.

COST RESPONSIBILITIES.   In establishing cost responsibilities, it is necessary to define exactly what resources are employed in serving each class of patron or in providing each type of service, and to identify the specific costs associated with each of these resources.

COST ALLOCATION.   After all elements of cost accruing at different functional locations have been classified according to their causes and cost responsibilities have been defined, total costs can be allocated to the various services in accordance with their particular cost responsibilities.

INCREMENTAL ANALYSIS OF UNIT COSTS—TO MAXIMIZE PROFIT WITHOUT MORE CAPITAL INVESTMENT

Incremental, or marginal, cost refers to the additional out-of-pocket cost incurred in producing each additional unit of output without an additional investment in production facilities. Rather than employing conventional accounting techniques in estimating and allocating overhead and other indirect loadings, only the actual cash flows are considered in incremental cost analysis. Thus, alternative levels of production are compared strictly on a cash basis.

In theory, as long as each unit of additional output generates revenues greater than its out-of-pocket cost, the increased production is profitable. The production level at which marginal cost is *equal* to marginal revenue returns the maximum profit.

Consider, for example, the situation described in Table 7.1. Up to 12 units of additional output can be produced without affecting the level of fixed costs, which remain at $150 throughout this range of alternatives. Total revenues are assumed to be a constant $150 per unit of output. Direct production costs can be expected to decrease slightly on a per-unit basis as the production rate increases, while other variable unit costs will probably increase. These other variable costs might represent additional promotional and advertising expenses, distribution costs, price discounts, or more liberal credit terms required to dispose of the additional production. In that such costs usually increase volume, their effect is equivalent to that of a price reduction.

As indicated in Table 7.1, and illustrated graphically in Figures 7.1 and 7.2, maximum profits will occur at a production level of between eight and nine units. This holds true whether or not fixed costs are included in the analysis, since only the *marginal* cost-*marginal* revenue relationship determines the optimum production rate.

This incremental cost concept plays an important role in the economic theory of cost and production; but as is the case with many important concepts in economic theory, its application to real-life problems is not always clear.

When applying the incremental cost concept, it is frequently necessary to stand back and take an overall view of the project being studied and its relationship to the company's entire operation. The use of incremental improvements to maximize profits for a given project or product may be in its best interests, but may fail to meet the company's profitability objectives or standards—in which case the costs associated with the higher production level should be diverted to other products that might generate a higher return. Since incremental cost studies are necessarily tied to past decisions, the time will come when the incrementally expandable project is best abandoned entirely. Whenever the project as a whole—including the considered incremental investment—fails to meet the company's goals, then that project exerts a detrimental effect on the company's total operation. The possibility of dropping the entire project should then be considered.

The incremental cost concept, however, illustrates an extremely important point in most engineering economy studies—economic evaluations should be based on the anticipated effects of the best *presently* available al-

Table 7.1 Example of Incremental Cost Analysis

Units of Production	Fixed Costs	Direct Production Cost	Other Variable Costs	Total Cost[a]	Total Revenue[b]	Total Profit[c]	Marginal Cost	Marginal Revenue
0	150	0	0	150	0	(150)	0	0
1	150	100	20	270	150	(120)	120	150
2	150	195	30	375	300	(75)	105	150
3	150	285	40	475	450	(25)	100	150
4	150	370	55	575	600	25	100	150
5	150	450	80	680	750	70	105	150
6	150	525	110	785	900	115	105	150
7	150	595	160	905	1050	145	120	150
8	150	660	225	1035	1200	165	130	150
9	150	720	320	1190	1350	160	155	150
10	150	775	450	1375	1500	125	185	150
11	150	825	640	1615	1650	35	240	150
12	150	870	900	1920	1800	(120)	305	150

[a] Fixed costs + direct production cost + other variable costs = total cost.
[b] Units of production × 150 = total revenue.
[c] Total revenue − total cost = total profit.

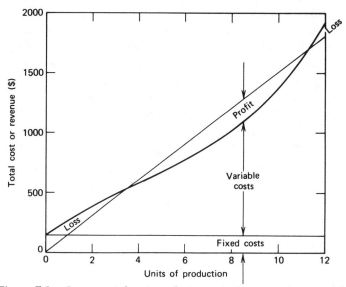

Figure 7.1. *Incremental cost analysis can point to maximum profit.*

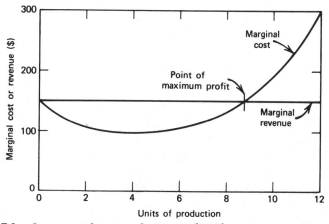

Figure 7.2. *Incremental cost analysis can identify optimum production level.*

ternative and not tied to bad decisions made in the past. The only costs that need be considered are those over which some degree of control can be exercised by present actions.

CASH AND CAPITAL BUDGETS COMPLEMENT PLANNING

Business operations are complex and difficult to forecast, which is sometimes used as an excuse for not operating a firm under close budgetary control. Complexity of operations, however, indicates the importance of emphasizing, rather than ignoring, the potential benefits of a thoughtfully prepared analysis of budgetary controls.

A budget is a "statement of expected revenues and expenditures." Budgetary control refers to an overall plan for analyzing and monitoring these expected revenues and expenditures by setting standards against which actual performance can be measured. When properly designed, the budget program can detect inefficiencies, fix responsibilities of individuals involved, and help insure the best possible utilization of the firm's physical and financial resources.

A proposed budget provides top management with the information needed to make intelligent, rational decisions. The effectiveness of an approved budget can be measured only in terms of its usefulness to the manager of the unit whose activities it covers.

One of the primary aims of a budgetary system is to aid in planning by complementing an action plan—a plan that will reveal the amount and timing of future resources needed to meet the firm's objectives. In this sense, a proposed budget can serve management as a short-term economic model, which can be used to evaluate the financial effects of proposed changes in objectives or operating procedures.

Another important advantage of coupling budgets with action plans is to provide a means of coordinating different activities within the firm. Each department can see exactly where and how it fits in with the firm's overall programs, what is expected of it, and the resources at its disposal. Potential jurisdictional conflicts can be identified and settled prior to their causing inefficiencies.

A budgetary program also provides an effective mechanism for controlling costs within the firm by assigning to specific individuals the responsibility for controlling designated segments of the budget and furnishing these individuals with guidelines for handling funds.

Finally, the budget program offers top management a broad perspective of the firm's entire operation, without requiring excessive involvement in operational nonproblem areas. That perspective allows management to relate the budget to the firm's short- and long-range plans. A comprehensive budget can serve as a financial blueprint of the firm's future.

The expenditure of any of the firm's resources can be budgeted; however, top management's primary concern is usually with the financial budget. Financial budgets can be broadly categorized as either cash and accrual budgets or capital budgets and may be either short range (1 year or less) or long range (usually from 1 to 10 years). The capital budget is primarily considered a planning tool, showing where the firm hopes to find itself on a given date in terms of a projected balance sheet. The cash budget, however, is an operational tool, indicating more finely tuned expectations or upper limits of expenses, rather than broad plans.

In both cash and capital budget analyses, the amounts should be as precise as possible, given the information available at the time of preparation. Naturally, the degree of precision varies inversely with the length of time over which the budget is projected. There should be at least enough detail to recognize variations from the expected values and to permit revisions whenever necessary.

CAPITAL BUDGETS.   The capital budget refers to estimated expenditures on capital account items—the firm's plans for replacing, improving, or acquiring income-producing capital assets. It is concerned with how much of, on what, and when the firm's capital or borrowed funds will be spent. Projected balance sheets constitute an important part of the capital budgeting process, reflecting anticipated changes in the firm's financial strength and providing a basis for evaluating financial effectiveness. Although not directly included in the capital budget, formal procedures for authorizing, analyzing, and reviewing all major capital expenditures are more profit-effective when tied in closely with the budgeting process.

CASH (AND ACCRUALS) BUDGETS.   The cash budget shows the projected flow of funds as related to the firm's operations. Cash budgeting involves the estimation, management, and control of these cash flows, summarizing anticipated receipts, disbursements, and accruals for a forthcoming period. Careful analysis of the cost figures included in the cash budget can reflect the cost efficiency of operations and the degree of realism in anticipated operating conditions. The cash budget serves as an operating aid by providing a fiscal early warning system. In some respects, the cash budget is to the firm's financial operations what standard costs are to anticipated factory production costs, in that actual performance is measured against predetermined standards. If those standards are realistic enough to be attainable, the result will be a measure of efficiency of production. If the standard is perfection, everyone is inefficient.

Useful budgets are developed in steps. There are five major steps involved in the analysis or preparation of a firm's budget program. The first four deal with developing the budget itself, and the fifth is concerned with its implementation.

BACKGROUND DATA.   The first step in preparing a budget is to accumulate data on the firm's past activities and operations. The information collected should include all pertinent financial data, as well as such operation indicators as the number, types, and sizes of projects conducted, clients, locations, and personnel utilization.

MARKETING ASSUMPTIONS.   The second step in preparing the budget is to develop basic assumptions regarding the components of the market for the firm's products or services as a whole. The market forecast in this step indicates the level of each component and thus the total volume of business from which the firm can derive its income.

THE COMPANY PLAN.   The next step in budget preparation is to develop basic assumptions regarding the firm itself, including its purposes, objectives, methods of operation, special business areas to be emphasized, amounts of time and money to be devoted to sales promotion and client contact, plans for acquiring new capabilities, and availability of and requirements for personnel, capital, and facilities. The results of this step tell how effectively the firm can compete in the markets defined in the previous step.

MAKING THE PROJECTIONS.   The final step in preparing the budget consists of making estimates of cash and capital flows for the future, based on a careful analysis of past performance, estimates of future operating performance, and assumptions about the economic and political environments in which the business will operate. The result should reflect management's best estimates of future sales, major expense items (such as staffing, salary and wage levels, and material costs), and any plans for expansion of facilities or other major capital commitments. The degree to which these projections approach eventual reality will vary directly with the degree and timeliness of involvement of the individuals responsible for carrying out each phase of the budgeted activities.

ACTION PROGRAM.   After the budget has been prepared and approved, each future decision must be weighed either to insure compatibility or to identify causes of variance with the projected figures. A continual review of all figures and assumptions included in the budget can generate appropriate revisions in light of changing conditions. If executives would view a budget as a means by which to measure their ability as money managers, they would see it as a practical and valuable tool, rather than as a means of limiting their accomplishments.

COST-BENEFIT ANALYSIS HELPS CHOOSE AMONG PUBLIC PROJECTS

The capital available for investment is always limited, and it is important to make sure that these funds are allocated and spent on a consistent and objective basis. This need for selectivity requires developing some measure of project worth to weigh against estimates of project cost. Project worth can be derived from cost-benefit and cost-effectiveness analyses. The cost-versus-benefit approach is becoming increasingly important in government, especially in federal projects. Standard procedures for cost-benefit analysis have been developed for use in water resources projects and defense planning and are equally desirable in recreation, highway, urban renewal, and public health programs.

Cost-benefit analysis offers several distinct advantages that cannot be obtained otherwise:

1. It focuses attention on key budget decisions, clearly identifying both immediate and long-range costs and consequences.
1. It identifies alternative ways of accomplishing goals with a given amount of money, or of using the available money for the maximum benefit.
3. It helps identify unprofitable or outdated programs or activities and provides the facts needed to make or support a decision to terminate them.
4. It enables management to establish and maintain control over several independent divisions or departments through control over both capital outlays and operations.

Cost-benefit analysis does not make decisions for management; it does make the costs and consequences of alternative actions clearer for the decision maker. By presenting the basic information on a proposed project, it forces a close examination of alternatives and eliminates any possible duplications.

The biggest problem in cost-benefit analysis is to measure objectively the benefits accruing from a proposed expenditure. It is often difficult even to foresee the consequences of a project, let alone to measure the values or benefits of these consequences in objective or quantitative terms. Even so, measurable costs can be weighed against measurable benefits; the *unmeasurable* benefits can then be examined to see if they might justify the measurable costs.

Social benefits are the most difficult to quantify. Nevertheless, cost-benefit analysis is especially important for projects in which investments that might not be financially attractive to private business are worthwhile from a social standpoint. These social considerations, in fact, largely dictate the conditions under which cost-benefit analysis is essential and explain why government-sponsored programs are particularly well-suited to this type of economic analysis.

There are three broad types of programs that must usually be government-supported and that require some assessment of intangible or hard-to-measure values. They deal primarily with peculiar conditions of consumption, production, and timing.

CONSUMPTION.   The usual procedure in a free-enterprise system is for the user to pay for the goods or services he consumes, the amount he is willing to pay indicating the value of the commodity to him. But charges for some commodities either are not collectible, or collection is difficult or impractical. For example, it is unwieldly to charge only the direct users for their use of police and fire protection, highways, national defense, outdoor recreation facilities, public health measures, and many municipal services. Since a collective benefit is created by these services, it may be socially or politically desirable to provide them, even though the amounts collectible from the direct users are not enough to cover their costs.

PRODUCTION.   Economies of scale may demand production facilities so large that private industry cannot possibly provide the necessary resources. Urban redevelopment, large hydroelectric projects, and highway systems are examples falling in this category.

TIMING.   In some projects, present costs are so high and payoffs are so distant that the short-term outlook of most private investors precludes consideration of projects involving such long-term benefits. For example, the present costs of recovering petroleum from shale do not necessarily reflect the importance of shale to future generations. Similarly, there is the possibility that long-term effects of urban renewal may ultimately far outweigh the immediate benefits.

The approaches to cost-benefit analysis are basically the same as the approaches used by private investors in evaluating investment proposals, the main difference being the means by which benefits are measured. The first step in the cost-benefit study is to project the physical output by years in whatever terms are appropriate—passenger miles or kilowatt hours, for example. Next, some estimate is made of the social value of these physical outputs, in dollars if possible. These two steps provide an estimate of the gross social contribution of the project. After they are completed, two popular methods of comparing costs and benefits are employed.

EQUIVALENT ANNUAL COST METHOD.   After the benefits have been computed for a typical or average year, the comparable operating and maintenance costs are computed for the same year. The initial capital costs are then converted to an equivalent annual cost basis by amortizing them over the expected life of the project. These annual capital costs, added to the annual operating and maintenance costs, give the total annual project cost. The ratio of the total annual cost to the gross annual benefits, then, gives the cost/benefit ratio to be used in comparisons with similarly derived ratios for other projects.

PRESENT VALUE METHOD.   In the second approach, the annual operating and maintenance costs for each year are deducted from the gross benefits for the same year, giving the annual net benefit. The resulting figures are then discounted back to the initial year, their total representing the present value of future benefits. The interest rate used in the discounting is that which the governing body must pay to finance the project with the sale of bonds. The ratio of the initial capital expenditure to this figure is the cost/benefit ratio.

The choice of method is not critical, since all have the same result if properly used. The important issues are to decide what benefits are to be included, how to value them, and what interest rates to use in amortizing the capital costs or discounting the operating costs involved.

While subjective measures of benefits are satisfactory for comparing alternative ways of accomplishing the same thing, objective measures are usually necessary to compare different programs having different objectives. An initial question to be answered is whether the benefits *can* be estimated reliably enough to even justify the expenditure for making the analysis.

Another, and possibly more significant, decision is whose costs and benefits merit inclusion in the analysis. For example, a municipal council might see a favorable benefit/cost ratio for an overpass because fewer traffic police would be required. Area motorists, on the other hand, might come to the opposite conclusion if they had to drive further on the way to and from work owing to the elimination of a grade crossing. Such a diversity can only be resolved by including and analyzing the costs and benefits for all parties.

Dollars naturally constitute the most convenient unit of measurement. One of the best clues in assessing the benefits of any project is always either the cost of providing an equivalent service through alternative means, or the cost of not providing the service at all.

Here are just a few ways that have been used to express benefits in dollars for different types of projects.

PUBLIC HEALTH.   The cost of curing or treating a disease serves as a measure of the benefits of preventing its occurrence. This measure is extremely conservative, since the value to an individual of not having some dreaded disease far exceeds the direct cost of this treatment.

URBAN RENEWAL.   The difference in property values before and after the elimination of blight and slums is certainly one direct monetary indicator of the benefits that accrue from such programs.

HIGHWAYS.   Decreased time and travel distance attributable to new highways are two of the many factors that might be translated into monetary terms as a direct measure of highway benefit.

RECREATION.   Some indication of the monetary value of public recreation facilities can be obtained by determining accurately what amount people are willing to pay elsewhere for the use of similar recreational opportunities.

TRAFFIC SAFETY.   The benefits of accident reduction can be expressed in terms of the monetary value of property and lives saved. Cost-benefit analysis can point out the specific areas where money could best be spent—such as in improved highways, more rigid law enforcement, motor vehicle inspection programs, and the like.

SUMMARY

Cost analysis provides accurate and factual information for management use in working toward a firm's goals and objectives by describing the firm's entire economic activity. Costs may be of a fixed, variable, or semivariable nature, depending on how they are related to the firm's capacity to do business and to its scale of operations. Several special fields of cost analysis are also important, including analysis of incremental costs, cash and capital budgets, and cost-benefit relationships. Incremental cost analysis is a useful tool in identifying optimum production levels or operating scales, whereas budget analysis serves as a valuable management planning aid. Cost-benefit analysis provides a quantitative basis for making decisions in relatively subjective or intangible areas—especially in connection with large government-supported programs conducted for the public good—which would otherwise have to be made almost intuitively. The cost analysis procedure involves four broad steps: (1) attributing costs to functions, (2) classifying costs by causes, (3) determining cost responsibilities, and (4) allocating costs according to responsibilities.

8

Cost Estimation

It is better to solve a problem with a crude approximation and know the answer, plus or minus 10 percent, than to demand an exact solution and not know the answer at all.

Author Unknown

Nothing improves the output of an engineering-economic study more than a good input; meaningful conclusions can be drawn only from meaningful data. Cost estimates usually constitute one of the main inputs for economic evaluations.

A variety of estimating methods is available to the engineer, depending on the purpose of his estimate. In general, there is a choice between three levels of detail and accuracy in cost estimating:

1. Order-of-magnitude estimates.
2. Semidetailed or conceptual estimates.
3. Detailed estimates.

Although there is no substitute for complete, detailed takeoffs and pricing when a final proposal is submitted for management approval at a guaranteed price (such as in a contractor's fixed-price bid), most feasibility

127

studies require far less sophistication. It is important, though, that the estimator be aware of the limitations of whatever estimating method he employs.

THE COST OF ESTIMATING RISES WITH PRECISION

The cost estimate can, within certain limits, be as precise as the engineer's company or client wishes, if they are willing to pay for it. The more accurate the estimate, the more time that is required for its preparation, and the more design data that must be developed prior to making the estimate.

Figure 8.1 shows the relative costs associated with the preparation of estimates having a specified level of precision, based on a $1 million project. Estimates on larger projects require a lower expenditure per project dollar, while smaller projects involve higher percentage estimating costs than shown here.

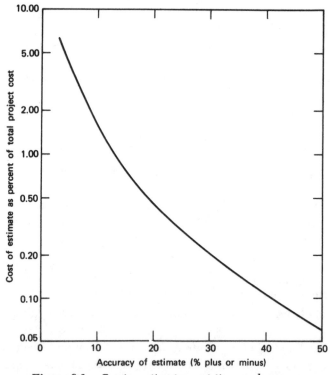

Figure 8.1. *Precise estimates cost time and money.*

For example, an estimate precise enough to justify a fixed-price bid— that is, within ±5 percent—would be about 4 percent of the total project cost, or $40,000 on a $1 million project.

An estimate of adequate precision for budgeting purposes—within ±10 percent—would cost 1.5 percent of the total project amount, or $15,000 in this case. An order-of-magnitude estimate, precise only to within ±30 percent, could be made for only 0.2 percent, or about $2000. For $600, a "seat-of-the-pants" or "ballpark" estimate, accurate within ±50 percent, could be had.

Ultimately, the actual cost of an estimate depends upon the estimator's skill and the availability and quality of the cost data with which he works.

AN ORDER-OF-MAGNITUDE ESTIMATE IS AN INEXPENSIVE STARTING POINT

The order-of-magnitude estimate offers a relatively low level of precision, varying by as much as 30–50 percent, and sometimes more, and seldom coming closer than within ±15 percent of the actual project cost. To achieve any better accuracy requires that the estimator be thoroughly familiar with the types of work under consideration and that he have access to pertinent cost data on similar projects. The sacrifice in precision, however, is often justified by the ability to screen a large number of alternative projects in a short time.

The value to be gained from short-cut estimates depends largely on the skill, experience, and judgment of the estimator. Estimates must be made with care and applied with caution and full recognition of their limitations. Even though considerable precision is sacrificed in the short-cut estimate, it can satisfy the need to quickly screen a wide range of alternatives. If warranted, the project in question can then be appraised in greater detail through more precise estimating methods.

Order-of-magnitude estimates can be based on cost-capacity relationships, ratios, or physical dimensions.

Cost-capacity relationships

One of the most useful short-cut estimating techniques—often used in chemical process industries and by cost engineers in other fields but seldom applied to civil engineering projects—employs the cost-capacity relationships among similar projects of different sizes. This method—sometimes called "factoring"—usually can be made to yield acceptable accuracy for study purposes.

The factoring method of short-cut cost estimating involves these steps:

1. Start with a recent project of known cost, similar in characteristics (but not necessarily similar in size) to the one under consideration.

2. Convert the cost of this previous project to a current basis, using an appropriate index to correct historical costs both for time and location. Some of the major general-purpose cost indexes used by estimators include the *Engineering News-Record Construction Cost and Building Cost Indexes,* the *Marshall and Stevens Installed Equipment Cost Index,* the *Nelson Refinery Construction Cost Index,* and the *Chemical Engineering Plant Construction Cost Index.* Also, numerous special-purpose cost indexes are compiled and published by various private firms and governmental agencies such as the Department of Commerce, Department of the Interior, Bureau of Public Roads, and Bureau of Reclamation. With the variety and availability of published indexes covering virtually all types of construction, the estimator should have little difficulty selecting an index appropriate to his interests.

Whatever index the estimator uses, the quality of his results requires him to know exactly what is included in that index. Important considerations include: frequency of publication, inclusion of labor productivity factors (to reflect improved productivity of field labor owing to advances in equipment and technology as well as design innovations in the facilities being constructed), and provisions for geographic differences in wage rates, materials costs, and labor productivity.

3. Define the relative size of the two projects in terms of resources consumed (gallons per day or kilowatts used, for example), end product units (number of classrooms, hospital beds, or apartment units), or physical dimensions (square feet, lineal feet, cubic feet, etc.).

4. Taking the three known quantities—the size and cost of the first project and the size of the second project—solve for the unknown cost of the second project by using an exponential power relationship:

$$\frac{C_A}{C_B} = \left(\frac{S_A}{S_B}\right)^x$$

where C_A represents the cost of project A, C_B is the cost of project B, S_A is the size of project A, and S_B is the size of project B, measured in the same units. Exponential x defines the cost-capacity relationship.

The exponential factor varies according to the type of project being considered, but usually is in the 0.6–0.8 range. However, factors as low as 0.2 and as high as 1.0 are not uncommon. Steam-electric generating plants, for example, generally have a cost-capacity factor of about 0.8. Waste treatment plants employing both primary and secondary treatment range between 0.7 and 0.8. Large public housing projects also average about 0.8. But steel storage tanks may have cost-capacity factors as low as 0.4 or as high as 0.8, depending on their shape.

An old estimating rule of thumb in manufacturing states that doubling the production increases costs by half; this represents an exponential factor of 0.585.

To illustrate how this short-cut "factoring" technique works, assume a 100-Mw steam generating plant was built in 1960 at a cost of $22 million, with the *ENR Construction Cost Index* at 800. The estimation of the present cost of a larger (350 Mw) but similar plant, with a corresponding cost index of 2000, involves:

Current cost of 100-Mw plant in 1983:

$$\$22 \text{ million} \times \left(\frac{2000}{800} \right) = \$55 \text{ million}$$

Current cost of 350-Mw plant:

$$\$55 \text{ million} \times \left(\frac{350}{100} \right)^{0.8} = \$55 \text{ million} \times 2.76$$

The estimated present cost of the 350-Mw plant, then, is about $150 million.

The development of appropriate cost-capacity factors for making short-cut cost estimates obviously requires the analysis of numerous completed projects. These data fortunately are available in several easily obtainable published sources. When the data are plotted on log-log graph paper, they usually show a straight-line relationship, reflecting this exponential economy of scale. The appropriate cost capacity factors can be calculated after fitting a straight line to the points. Table 8.1 shows typical cost-capacity factors for a variety of plants and facilities.

As with any other short-cut method, factoring can result in sizable errors if the estimator is not familiar with the project under consideration or with

Table 8.1   Cost-Capacity Factors Vary by a Factor of 1.5/1 or Less

Type of Facility	Cost-Capacity Factor	Units of Capacity
Acetylene plant	0.73	tons/day
Aluminum plant	0.76	tons/year
Ammonia plant	0.72	tons/day
Steam boiler plant	0.75	pounds/hour
Cement plant	0.86	tons/day
Chlorine plant	0.62	tons/day
Electric generating plant (nuclear)	0.68	megawatts
Electric generating plant (steam)	0.79	megawatts
Industrial building	0.67	square feet
Municipal incincerator	0.80	tons/day
Oxygen plant	0.72	tons/day
Public housing project	0.75	number of rooms
Refrigeration system (mechanical)	0.70	tons
Sewage treatment plant (primary only)	0.68	gallons/day
Sewage treatment plant (primary and secondary)	0.75	gallons/day
Storage tank	0.63	gallons
Sulfuric acid plant	0.67	tons/day
Utility distribution main (gas and water)	0.91	pipe diameter
Utility distribution main (gas and water)	0.82	length installed

the inherent limitations of the technique. Most significant is the selection of an appropriate exponent for factoring. For example, if an exponent of 0.7 were used in estimating the cost of a proposed project having five times the capacity of a previous one, when the correct exponent is actually 0.8, an error of about 17 percent in the final estimate would result. If the difference in project size were greater, the error also would be greater; a nearly 26-percent error would be associated with a 10:1 ratio of project sizes.

Equipment ratios, physical dimensions, or weight can also provide a basis for factoring. The cost of projects involving a few major items of equipment that can be accurately priced is estimated by applying a multiplier to the cost of these items. Ratio estimates are often used in chemical process industries in which specialized equipment makes up the major portion of the total project cost.

Physical dimensions or weight can also be used to estimate the approximate cost of many different types of projects. Building cost estimates, for example, can be made in terms of square feet of floor space or cubic feet of building volume, while concrete work can be priced on a per-cubic-yard basis.

Equipment costs can sometimes be estimated with surprising accuracy simply on the basis of their weight. For this purpose, equipment can be grouped into three broad categories: (1) precision, (2) mechanical/electrical, and (3) functional.

Precision equipment may be electronic or optical and generally costs about 10 times as much per pound as mechanical/electrical equipment, which in turn costs about 10 times as much per pound as functional items. In the functional class, items such as large power tools, automobiles, engine-generator sets, and heavy construction equipment all cost approximately the same on a per-pound basis. Thus, the costs of a 50,000-lb motor grader, a 70,000-lb crawler tractor, and a 90,000-lb pipelayer can all be estimated with reasonable accuracy (within 30 percent) by comparison with the cost per pound of an ordinary passenger automobile. Similarly, the cost of mechanical equipment—where a large part of the product's weight is made up of either electrical or mechanical parts—does not vary too much from the cost per pound of conventional small home kitchen appliances. To acquire a rough feel for precision equipment costs, the estimator need only check the cost per pound of items such as cameras and small electronic testing equipment.

In all types of order-of-magnitude estimates, the accuracy achieved is largely a function of the estimator's judgment and experience; he must first be able to visualize the physical fulfillment of the project. Any projects identified as being potentially favorable can then be appraised in greater detail through more precise estimating methods.

SEMIDETAILED ESTIMATES ARE USUALLY SUFFICIENT

Semidetailed, conceptual, or budget estimates, should be accurate to within about 10 percent of the most precise estimate possible. This level of precision is usually adequate for making decisions regarding project feasibility—whether or not the owner decides to proceed with the project. The engineer's estimate on new construction work might be considered a semidetailed estimate in most cases. The precision of the semidetailed estimate depends on the amount and quality of information available at the time the estimate is made.

More information is required for making semidetailed estimates than for making order-of-magnitude estimates. Instead of using mathematical relationships between historical costs and estimated costs on proposed work, the new project must be considered on its own. Actual quotations are obtained on major equipment and material items; some design data are necessary for making rough takeoffs; and approximate unit costs may be applied to the measured units.

The semidetailed estimate can be used to advantage by the engineer in several ways. Since the precision is generally sufficient to warrant the authorization of a project by an owner, the semidetailed estimate may be all that is needed, so long as the owner is aware of its limitations. This type of estimate is also useful in providing a rough check on detailed estimates obtained through more refined methods.

DETAILED ESTIMATES MAXIMIZE PRECISION BUT NOT NECESSARILY ACCURACY

Detailed estimates are used as a basis for making bids. These estimates should be precise within 5 percent, having been prepared from complete engineering specifications, drawings, and site surveys. They are, however, time-consuming, costly to prepare, and merited only when absolutely necessary. In many cases, the contractor is the only one capable of making a valid detailed estimate.

Considerable information is required for a detailed estimate, often more than is available at the time the engineer's estimate must be made. In preparing the estimate, the specifications and plans are studied carefully, quantity takeoffs are made, prices are obtained, labor availability and wage rates are checked, subcontractors' estimates are requested, and schedules are made up. There are, in every detailed estimate, many opportunities to make mistakes, usually resulting in estimates that are too low. The most common errors include omission of items, undermeasurement of quantities, and underestimation of labor requirements.

Precision does not guarantee accuracy. In fact, the actual costs are usually higher than the most precise estimates.

CAPITAL VERSUS OPERATING COSTS

Typically, an engineering economy study compares a high-investment, low-operating-cost project with a low-investment, high-operating-cost alternative. Therefore it is important that both capital and operating costs be estimated with the same degree of precision.

The three basic types of estimating methods can be applied equally well to both capital and operating costs, their applicability depending on the purpose of the study. Either the order-of-magnitude or the semidetailed budget estimate is probably adequate for most engineering economy studies.

CAPITAL COST ESTIMATES.   The total capital requirements of a new manufacturing plant can be broken down into the following components for estimating purposes.

1. Depreciable investment.
   a. Buildings and utilities.
   b. Equipment, including installation.
   c. Other.

2. Expensed or amortized investment.
   a. Research and development.
   b. Engineering.
   c. Startup.
   d. Other.

3. Nondepreciable capital requirements.
   a. Land.
   b. Working capital.
      i. Cash.
      ii. Receivables.
      iii. Inventory.

Additional breakdowns of these items can be made if the results justify further detail. But unless there is some compelling reason to introduce complexity into the study, a relatively simple and straightforward treatment is preferable.

OPERATING COST ESTIMATES.   Operating costs, or in the case of an industrial plant, production or manufacturing costs, can be divided into two broad categories: unit costs and period, or time, costs.

1. Unit costs.
    a. Materials.
    b. Labor.
    c. Variable overheads.
2. Period (time) costs.
    a. Fixed overheads.
    b. Capital charges.

Unit costs are estimated on a cost per unit of production or output basis and are generally linear over a wide range of production volume. Period costs are incurred at a fixed level over a prescribed range of output, since they are related to the level of investment rather than to the level of production.

A different type of cost breakdown is necessary in the case of a commercial, as opposed to an industrial, venture; but the *types* of costs—unit and period or fixed and variable—are basically the same.

HOW TO MAKE QUICK MANUFACTURING COST ESTIMATES

There are many instances in which a high level of precision is neither necessary nor possible in manufacturing cost estimates and in which the available data are useful only over narrow ranges or are of doubtful reliability. Sometimes no data are available.

In these situations, any estimate may be considered better than none, and the quickest, roughest figures are gratefully accepted. The approach described here provides such figures.

SOURCES OF DATA.   When no data are available that pertain directly to the project in question, the engineer must turn to the literature for his economic information. Fortunately, a considerable amount of financial and operating data is available for all types of manufacturing industries from readily accessible published sources. For example, the U.S. Department of Commerce publishes annually its *Survey of Manufactures*, which gives general statistics for several hundred different industries and industry groups. The data reported in the survey cover labor, materials, and other costs, as well as the total value of shipments for each industry. These data are particularly useful as a guide to the overall operating cost structure of manufacturing industries.

Other sources, such as Prentice-Hall, *News Front*, and *Dun & Bradstreet*, tabulate and publish industry data from the financial records of major U.S. corporations. They summarize sales, assets, net worth, profit, and other useful financial data both by industry gorup and by individual companies.

If a new venture is being planned in some unfamiliar manufacturing line, a logical first step in the economic evaluation is to examine the financial and operating characteristics of individual companies involved in that field of manufacturing. Putting together information that is easily obtained from published sources can quickly give a reasonably clear picture of what a new venture might expect in any given field.

In making quick manufacturing cost estimates, all costs are related to annual sales. After the approximate scale of operations has been established, the expected annual sales can be estimated. For example, if a new plant were being considered to produce crushed stone for highway use at a rate of 500,000 tons per year, and if the stone would sell for $5.00 per ton, then the plant would generate net sales of $2,500,000 annually. From just this much information, a preliminary economic picture of the operation can be drawn.

CAPITAL REQUIREMENT.   Table 8.2 shows the total assets and net worth per dollar of annual sales for the 20 major manufacturing industry groups. Crushed stone falls in Standard Industrial Classification (SIC) category 32—stone, clay, and glass products—in which the total investment is in the neighborhood of $1.00/dollar of annual sales. If the proposed crushed stone plant were assumed to be typical of that industry group, the total investment would be in the same neighborhood, or about $2,500,000.

PRODUCTION COSTS.   Table 8.3 summarizes the breakdown of production costs on a per-dollar-of-sales basis for the same 20 manufacturing industry groups. It can be seen that direct labor and materials costs account for 62 percent of the total sales dollar, while indirect payroll costs come to 8 percent of sales. This leaves 30 percent to cover fixed costs such as depreciation, interest, insurance, property taxes, and general administration expenses. Table 8.4 summarizes the operating economics of the proposed crushed stone plant, again using the SIC-32 industry averages.

As shown in Table 8.4, direct costs of $1,550,000 for labor and materials leave $950,000 to cover fixed costs. Deducting administrative or indirect payroll expenses of $200,000 leaves $750,000.

Table 8.2   Financial Characteristics of Major Industry Groups

SIC	Industry Group	Average Economic Life of Equipment (years)	Ratio Total Assets per Net Sales	Net Worth per Dollar of Annual Sales ($)
20	Beverages	12	0.48	0.27
21	Tobacco	15	0.69	0.38
22	Textile mill products	14	1.04	0.54
23	Apparel	9	0.62	0.29
24	Lumber and wood products	10	0.99	0.54
25	Furniture and fixtures	10	0.63	0.41
26	Paper products	14	0.99	0.54
27	Printing and publishing	11	0.84	0.46
28	Chemicals	11	0.89	0.52
29	Petroleum products	16	1.18	0.72
30	Rubber and plastics	12	0.84	0.40
31	Leather	11	0.56	0.30
32	Stone, clay, and glass	15	1.00	0.61
33	Primary metals	16	1.15	0.64
34	Fabricated metal products	12	0.72	0.40
35	Nonelectrical machinery	12	0.90	0.52
36	Electrical equipment	10	0.69	0.35
37	Transportation equipment	12	0.63	0.33
38	Instruments	12	0.85	0.54
39	Miscellaneous manufacturing	12	0.70	0.43
	All manufacturing (average)	12	0.85	0.46

Fixed costs are related primarily to fixed investment, with depreciation and interest the most important elements. The average economic life of equipment in SIC-32 industries is 15 years (see Table 8.2), well above the allowable recovery rate of 5 years, as established by the tax legislation of 1981. Average interest costs are about 1.5 percent of the total investment, depending on the firm's cost of capital and on the amount and method of financing employed.

Insurance and property taxes are generally about 1.5 percent of the fixed investment, and miscellaneous administrative and general investment-related expenses can be included at 1.0 percent annually.

In the example, fixed costs total $800,000, leaving a net profit before

Table 8.3   Operating Characteristics of Major Industry Groups

SIC	Industry Group	Direct Labor ($)	Direct Materials ($)	Indirect Payroll ($)	All Other ($)
		Cost per Dollar of Annual Sales			
20	Beverages	0.08	0.65	0.05	0.22
21	Tobacco	0.07	0.55	0.02	0.36
22	Textile mill products	0.20	0.56	0.05	0.19
23	Apparel	0.23	0.48	0.07	0.22
24	Lumber and wood products	0.21	0.52	0.05	0.22
25	Furniture and fixtures	0.24	0.45	0.09	0.22
26	Paper products	0.17	0.51	0.07	0.25
27	Printing and publishing	0.20	0.33	0.16	0.31
28	Chemicals	0.09	0.43	0.07	0.41
29	Petroleum products	0.04	0.75	0.02	0.19
30	Rubber and plastics	0.20	0.45	0.08	0.27
31	Leather	0.25	0.46	0.07	0.22
32	Stone, clay, and glass	0.21	0.41	0.08	0.30
33	Primary metals	0.17	0.56	0.06	0.21
34	Fabricated metal products	0.21	0.46	0.09	0.24
35	Nonelectrical machinery	0.21	0.42	0.12	0.25
36	Electrical equipment	0.19	0.42	0.14	0.25
37	Transportation equipment	0.15	0.57	0.08	0.20
38	Instruments	0.17	0.34	0.14	0.35
39	Miscellaneous manufacturing	0.21	0.44	0.10	0.25
	All manufacturing (average)	0.16	0.51	0.10	0.23

taxes of $150,000. Deducting income taxes at a 50-percent rate results in a net profit after taxes of $75,000. Adding the depreciation gives a net annual cash flow of $575,000, or 23.0 percent of the total investment. Referring to interest tables, it is found that a capital recovery rate of 23.0 percent (or 0.230) represents a net return of about 22 percent annually over a 15-year period, the estimated economic life of the project. Since the cost of capital

Table 8.4   Example of Preliminary Cost Estimate for Crushed Stone Operation

Capacity 500,000 tons/year
Net product price $5.00/ton

Net annual sales: 500,000 tons at $5.00/ton	$2,500,000
Total investment	2,500,000
Direction production costs at 62% of annual sales	1,550,000
Gross operating profit	$ 950,000
Indirect costs	
Indirect payroll at 8% of annual sales	$ 200,000
Average depreciation at 20% of investment	500,000
Average interest at 1.5% of investment	37,500
Other indirect expenses at 2.5% of investment	62,500
Total indirect costs	$ 800,000
Net profit before income taxes	$ 150,000
Income taxes at 50% of pretax net profit	75,000
Net profit after taxes	$ 75,000
Plus depreciation	500,000
Net annual cash flow	$ 575,000

is likely to be considerably less than the 22 percent that the project is expected to generate, the proposed venture could be considered—at this early stage—economically feasible.

USEFULNESS OF THE ESTIMATES.   Obviously, many refinements can be made in this rough estimate as better data become available. A far truer picture can be drawn even from published literature. The important point is simply that a rough but useful estimate can be made quickly and easily, without any specific data, to test the reasonableness of a proposed manufacturing operation. When better information becomes available, it can replace the general industry standard figures until a reasonably accurate estimate is ultimately obtained.

ESTIMATING THE ACCURACY OF ESTIMATED COSTS

A lot of money is either spent or committed on the basis of an engineer's estimate. Often the engineer who made the estimate can defend its degree of precision but is uncertain about its accuracy.

The engineer's estimate is usually made up of several uncertain and only loosely predictable elements. These elements are either multiplied or added together, or both, to come up with a final figure. Under such conditions, it is surprising that most engineer's estimates are as accurate as they are.

The next best thing to being certain about an estimate's accuracy is to be certain about just how uncertain the estimate is. Adding numbers together increases the accuracy of their total, whereas the other arithmetic operations—subtraction, multiplication, and division—magnify any inaccuracies present in the individual numbers. If enough separate items can be added together, fairly good accuracy can often be attained without being especially precise on individual items.

CONTRACTOR'S BIDS VERSUS ENGINEER'S ESTIMATES.   A comparison of engineer's cost estimates and actual contract award prices was made on nearly 100 recent projects, drawing from those reported over the past year in the "Unit Prices" section of *Engineering News-Record*. The jobs ranged in size from $119,000 to more than $15 million and involved from 2 to 13 bids. The results are summarized in Table 8.5.

Only 3 percent of the engineer's estimates were more than 20 percent below the low bidder's price; similarly, just 3 percent of the estimates were off by more than 20 percent on the high side. Some 30 percent of the engineer's estimates were within 5 percent of the actual contract award price; just over half were within 10 percent; about three-fourths were within 15 percent; and 94 percent of the estimates came within ±20 percent.

Table 8.5   Engineer's Estimates can Approximate Contractor's Bids

Ratio of Low Bid to Engineer's Estimate	Percent of Contracts Awarded
0.800 or less	3.0
0.801–0.850	15.2
0.851–0.900	19.7
0.901–0.950	10.6
0.951–1.000	15.2
1.001–1.050	15.2
1.051–1.100	10.6
1.101–1.150	4.5
1.151–1.200	3.0
1.201 or more	3.0
	100.0

ACCURACY OF PRODUCTS.   When two or more independent variables are multiplied together, any inaccuracies in the individual variables are amplified in their product. The mathematical expression for a two-variable relationship, say range $(A \times B)$, is:

$$\sqrt{A^2 \times \text{range}^2(B) + B^2 \times \text{range}^2(A)}$$

where the range of the variable is expressed in absolute units, not as a percent. For example,

$$(100 \pm 10\%) \times (60 \pm 20\%)$$

$$= (100 \pm 10) \times (60 \pm 12)$$

$$= (100 \times 60) \pm \sqrt{100^2 \times 12^2 + 60^2 \times 10^2}$$

$$= 6000 \pm \sqrt{1,440,000 + 360,000}$$

$$= 6000 \pm 1340$$

$$= 6000 \pm 22.4\%$$

Table 8.6 shows the percent range that can be expected in the product of two independent variables, depending on their respective accuracies (or inaccuracies).

Table 8.7 illustrates how the information in Table 8.6 is used in a typical engineering cost estimate. For example, in the rock excavation estimate, the quantity (accurate within $\pm 15\%$) is multiplied by the unit cost (accurate within $\pm 30\%$) to arrive at the estimated total cost for that item, which is accurate within ± 33.6 percent (from Table 8.6).

Division works exactly the same way as multiplication—increasing the

Table 8.6   Percent Range Expected in Product of Two Independent Variables

Percent Range of Variable A	Percent Range of Variable B						
	0	5	10	15	20	25	30
0	0	5.0	10.0	15.0	20.0	25.0	30.0
5	5.0	7.1	11.2	15.8	20.6	25.5	30.4
10	10.0	11.2	14.1	18.0	22.4	27.0	31.6
15	15.0	15.8	18.0	21.2	25.0	29.2	33.6
20	20.0	20.6	22.4	25.0	28.3	32.1	36.1
25	25.0	25.5	27.0	29.2	32.1	35.4	39.1
30	30.0	30.4	31.6	33.6	36.1	39.1	42.5

Table 8.7   Accuracy of the Product of Several Inaccurate Figures

Item	Estimated Quantity	Estimated Unit Cost	Estimated Total Cost
Rock Excavation	6,000 CY ± 15%	$5.75 ± 30%	$34,500 ± 33.6%
Common excavation	15,000 CY ± 10%	$2.75 ± 30%	$41,250 ± 31.6%
Embankment	30,000 CY ± 10%	$1.50 ± 25%	$45,000 ± 27.0%
Cement	35,000 bbls ± 5%	$4.90 ± 10%	$171,500 ± 11.2%
Reinforcing steel	50 tons ± 5%	$325.00 ± 5%	$16,250 ± 7.1%

range of inaccuracy. In multiplication or division, the answer is less accurate than the least accurate of the individual numbers.

ACCURACY OF SUMS.   Fortunately, the accuracy lost in multiplying and dividing can often be regained in adding. While multiplication and division decrease accuracy, addition improves accuracy. Mathematically,

$$\text{range}\,(A + B) = \sqrt{\text{range}^2\,(A) + \text{range}^2\,(B)}$$

where the range is again expressed in absolute units. Another example is:

$$(100 \pm 10) + (60 \pm 12)$$
$$= 160 \pm \sqrt{10^2 + 12^2}$$
$$= 160 \pm 15.6$$
$$= 160 \pm 9.8\%$$

In this case, the expected error range of the total is less than the error in either of the individual numbers.

If one number were subtracted from the other, the expected range (in units) would be the same as if they were added, but the percentage range would be far greater. Using the same example:

$$(100 \pm 10) - (60 \pm 12)$$
$$= 40 \pm 15.6$$
$$= 40 \pm 39\%$$

Table 8.8 shows how the accuracy of the sum of the five estimated items from Table 8.7 is improved over their individual accuracies. Adding the squares of the five dollar items having individual accuracy ranges of ±33.6,

Table 8.8 The Whole Is More Accurate than the Parts

Item	Estimated Cost ($)	Accuracy of Estimate Percent	Accuracy of Estimate Dollars	(Dollar Range)^2
Rock excavation	34,500	± 33.6	± 11,600	124,600,000
Common excavation	41,250	± 31.5	± 13,040	170,000,000
Embankment	45,000	± 27.0	± 12,150	147,600,000
Cement	171,500	± 11.2	± 19,210	369,000,000
Reinforcing steel	16,250	± 7.1	± 1,150	1,320,000
Total	308,500	± 9.2	± 28,300	812,520,000

31.6, 27.0, 11.2, and 7.1 percent and taking the square root of that sum results in a 9.2 percent accuracy of the total cost of the five components. This "principle of compensating errors" largely explains the accuracy of engineers' estimates, even when accurate estimating data are not available at the time the estimates are made.

SUMMARY

Cost estimates provide one of the most important inputs for economic evaluations. Although the precision of an economic analysis can be no better than the accuracy of the cost data used, there are many situations in which a high level of precision is neither necessary nor possible; short-cut estimating methods may be used to advantage in such cases. Order-of-magnitude estimates, accurate to within ± 30 percent, are especially useful in preliminary studies; they may be based on historical cost-capacity relationships or utilize equipment ratios, physical dimensions, or weight as the basis for estimating costs. Semidetailed estimates are used in making budget decisions and should be accurate to within about 10 percent. The detailed estimate, accurate within 5 percent, is usually necessary only for contractors making fixed-price bids. In any cost estimate, the capital and operating costs should be established with comparable precision, with the degree of detail in the cost breakdown depending on the purpose of the estimate. For the very quickest and roughest cost estimates on new ventures in which specific information is unavailable, general industry economic data can be used initially and refined later, as better data become available. When the individual numbers in an estimate are relatively inaccurate, arithmetic operations such as subtraction, multiplication, and division increase the inaccuracies. However, adding numbers together improves the accuracy of their total, and fairly good overall precision can be obtained in an estimate made up of a large number of relatively inaccurate individual items.

9

Equivalent
Annual Costs

*Gain may be temporary and uncertain; but ever while you live, expense
is constant and certain.*

Benjamin Franklin

All businesses incur two main types of cost: fixed and variable. Fixed costs
are associated with the firm's capacity to do business, are related to its capital investment, and accrue with time rather than with the level of operation. Variable costs increase or decrease more or less directly with the firm's
output of goods or services.

INDUSTRY COST STRUCTURES REFLECT INVESTMENT
IN FIXED ASSETS

Some industries are characterized by a high capital investment in fixed
assets. Other industries operate on relatively little fixed capital, but usually
experience a high variable cost rate. Table 9.1 shows the capital require-

ments and fixed-variable cost relationships typical of several selected major industry groups.

The rail and motor freight transportation fields offer a good example of two competing fields that operate with almost opposite capital-operating cost structures. Railroads have a high investment per dollar of sales ($2.68); truck lines have a far lower investment requirement in fixed assets and a correspondingly lower fixed cost. About 60 percent of railroad revenues go to cover fixed costs, as compared with 29 percent of the truckers' income. Direct operating costs for railroads, however, are far lower than for truck lines—40 percent of the sales dollar, as opposed to 71 percent. Hence, railroad schedules must reflect high fixed and low operating costs. Truck shipment rates, however, are structured to reflect low fixed costs and high operating costs.

Other examples of high-fixed-cost, low-variable-cost industries include: electric and gas utilities, with more than $3.00 invested for each dollar earned annually; telephone utilities, with a total investment of 2.7 times their annual revenues; and airlines and mining companies, whose average investment is $1.40 per annual sales dollar.

Table 9.1   Cost Structure of Selected Industry Groups

Industry	Total Assets as Percent of Annual Sales	Cost as Percent of Sales	
		Fixed	Variable
Agriculture	89	33	67
Mining	145	42	58
Construction	83	32	68
Manufacturing	85	33	67
Railroad transportation	268	60	40
Motor freight transportation	57	29	71
Air transportation	138	41	59
Telephone and telegraph	272	61	39
Public utilities	318	68	32
Wholesale trade	42	26	74
Retail trade	58	29	71
Food stores	20	23	77
Business services	89	33	67
Engineering and architectural services	43	26	74

Having a high level of fixed costs means that a firm must maintain a consistently high volume or utilization rate to operate successfully. A freight train or airplane incurs about the same costs over a given distance, whether it travels full or half full, and a public utility's total costs vary little, regardless of the cubic feet of gas or kilowatt-hours of electrical energy sold. In all these cases, sales receipts or revenues vary almost directly with volume.

In low-capital-cost, high-variable-cost industries, the opposite is true. If most costs are of a variable nature, reduced sales can be met with cutbacks in production, which, in turn, will be reflected in lower costs.

At high levels of production, fixed-cost-oriented industries have a distinct economic advantage over their variable-cost-oriented competitors. During recessions, on the other hand, high-variable-cost-structured firms have an advantage because of their ability to cut costs.

For these reasons, high capital cost structures are usually associated with industries or firms whose products or services benefit from stable, or at least predictable, demands.

CAPITAL IS CHEAPER THAN LABOR—UP TO A POINT

Even within a single industry, management has many options regarding the firm's capital-operating (or fixed versus variable) cost structure. Over a wide range of operations, capital can be substituted for labor, thus substituting fixed costs for variable costs. In the extreme case in which capital assets replace *all* production labor, a firm operates with fixed costs only, exclusive of raw materials. Its costs of doing business in this way are, therefore, almost uniform from year to year. This has almost been accomplished in the petroleum refining field, in which direct labor costs now account for less than 5 percent of total sales; raw materials are the only important variable costs remaining.

Regardless of the industry or type of business, a $5 capital investment can reduce direct labor costs by about $1. This relationship partially explains why the substitution of capital for labor is not necessarily desirable from the company's standpoint in many industries. If higher variable costs can be compensated for by higher prices, as is often the case, then it may be to a firm's advantage to retain the operating flexibility associated with a high variable cost rate and a minimum of fixed charges. The same is true in industries in which technological change occurs rapidly and is accompanied by asset obsolescence. Even in fields in which high overheads might not be objectionable because of lower direct production or operating costs, the annual fixed charges on capital—typically in the 15–20 percent range—mean that little or no real economy is attained by the capital-labor substitution.

A realistic evaluation of the relative merits of various capital-operating cost alternatives can result from expressing both capital and operating costs on an equivalent annual basis. A comparison of equivalent annual costs can quickly establish the optimum cost structure for an industry, firm, or new venture.

THE COST OF A CAPITAL INVESTMENT CAN BE EQUATED TO ANNUAL PAYMENTS

Most engineering economy studies identify a preferred choice between or among alternatives. Since most engineering proposals involve expenditures and receipts of money over a period of time, an equitable basis is needed for comparing alternatives whenever the time periods or amounts differ. A convenient way to achieve such a basis is by taking all future amounts and converting them to a single figure—the equivalent annual cost.

The equivalent annual cost of a time series of unequal payments is the uniform amount that would have to be set aside annually to have exactly the same economic effect as the unequal series of payments. Stated another way, it represents an annuity whose present value is equal to that of the unequal payments over the same time period and at the same interest rate.

For example, an individual with a $60,000, 20-year, 15-percent mortgage on his home will spend $798 monthly over the 20-year period to cover the principal and interest on this loan. This $798 monthly payment for 20 years is, economically speaking, equivalent to a $60,000 cash outlay now, or to a single $982,000 payment 20 years from now, or to a $9586 annual payment made at the end of each of the 20 years. These four alternatives all have exactly the same present value—$60,000—and the $9586 figure is said to be the equivalent annual cost of the mortgage.

The equivalence of these four spending plans is not immediately apparent from the absolute figures alone; for equivalence depends not only on the amounts involved but also on the timing of those amounts and on the interest or discount rate. A change in any one of these three factors changes the equivalence relationships.

EQUIVALENT ANNUAL COSTS CAN EQUATE DIFFERING COMPONENTS

All types of firms find themselves choosing among assets having different initial costs and offering different periods of useful service. Problems of equivalence, for example, arise frequently in feasibility studies concerning public utilities in which the company's financial structure and regulatory controls impose restrictions on how capital expenditures can be charged. Many of these problems can be quickly and easily resolved on an equivalent annual cost basis.

For example, capital charges refer to costs that are directly attributable to the amount of capital invested and are largely independent of the amount of production or sales. The annual charges against a firm's capital assets may include six major items: depreciation, interest, *ad valorem* taxes, administrative and general expenses, return on assets, and income taxes. Some of these factors may be omitted or included elsewhere in the company's accounts.

Depreciation, the largest of the six, represents the annual cost of recovering a capital expenditure through accounting deductions and is allocated in the maximum annual amounts allowed by tax regulations. Depreciation is usually computed as a fixed percentage of the initial cost (straight-line method), or as a varying percentage, as defined by the IRS in its accelerated recovery tables.

Interest, like depreciation, varies directly with the investment. If the investment is financed through outside borrowing, interest charges are included directly as a capital charge. If the investment is internally financed, interest charges may show up instead as a cost of capital. The capital recovery rate or factor, often used in engineering economy studies, includes both depreciation and interest in a single figure.

Ad valorem taxes, or property taxes, are paid by utilities and other firms on the value of their property. Although the tax rate and the tax base may vary from year to year, *ad valorem* taxes are generally computed as a percentage of the original cost of the property less accrued depreciation (also called the net cost or book value) or as a percentage of some "assessed" value.

General and administrative (G & A) expenses, typically accounting to between 1.0 and 2.0 percent of the total investment, refer primarily to the expenses involved in acquiring the asset initially and in keeping track of it over its economic life. General and administrative expenses are sometimes lumped in with general overheads or fixed costs, rather than being charged against specific assets.

Return on investment (ROI) is the annual cost of money assessed against a project, whether financed by equity, debt capital, or a combination of the two. Return for a utility firm is usually expressed as a percentage of the original cost, original cost less accrued depreciation, fair value, net worth, or some other rate-based standard. For industrial firms, the return can probably be expressed as a percentage of total investment or net worth. Return represents the cost of the use of money, as opposed to depreciation expense, which refers only to the tax-regulated recovery of the initial out-of-pocket acquisition cost.

Income taxes, both federal and state, are calculated from the firm's profits; thus, they are indirectly related to the same base as the return. A firm in a 50-percent tax bracket, for example, requires a gross pretax return of twice its projected after-tax net profit to allow for payment of income tax.

The actual amounts of these capital changes may vary widely among different industries and in different regions of the country. Nevertheless, their effect is essentially the same in all instances.

ALTERNATIVES EQUATE THROUGH ANNUAL COSTS

In the following example, an asset with an initial cost of $10,000 is to be depreciated, using accelerated recovery, over a 5-year period. The other capital charges—property taxes, ROI, and income taxes—are assigned percentage rates of 2.0, 15.0, and 12.5, respectively, applied only against the average undepreciated balance each year. Table 9.2 summarizes the capital costs by year over the 5-year period.

The problem, then, is to convert this unequal 5-year flow of costs to an equivalent single figure that can be used as a basis for comparison with investment alternatives. This is done by taking the present value of the total annual cost of each year, using as a discount rate the company's net ROI (15.0 percent). The results are given in Table 9.3.

Dividing the total present value of annual costs ($11,816) by the sum of the discount factors (3.353) gives the equivalent annual cost of the series of payments—$3524. This figure is verified by the calculations in Table 9.4.

The net economic effect—that is, the present value of all future costs—is the same in both cases. The equivalent annual cost figure can, therefore, be

Table 9.2 Annual Costs for an Income-Generating Asset ($10,000 First Cost and 5-Year Life)

Year	Account Balance (R)			Recovery Charges	Annual Charges on Capital			Total Annual Cost
	January 1	December 31	Average Balance		Property Taxes (2%)	ROI (15%)	Income Taxes (12.5%)	
1	10,000	8,000	9,000	2,000	180	1,350	1,125	4,655
2	8,000	4,800	6,400	3,200	128	960	800	5,088
3	4,800	2,400	3,600	2,400	72	540	450	3,462
4	2,400	800	1,600	1,600	32	240	200	2,072
5	800	0	400	800	8	60	50	918

Table 9.3   Present Value of Annual Costs for a $10,000 Asset with 5-Year Life

Year	Total Annual Cost ($)	×	Discount Factor (15%)	=	Present Value of Annual Cost ($)
1	4,655		0.870		4,050
2	5,088		0.756		3,847
3	3,462		0.658		2,278
4	2,072		0.572		1,185
5	918		0.497		456
Average	3,524	=	3.353		divided into 11,816

Table 9.4   Equivalent Annual Costs for a $10,000 Asset with 5-Year Life

Year	Total Annual Cost ($)	×	Discount Factor (15%)	=	Present Value of Annual Cost ($)
1	3,524		0.870		3,066
2	3,524		0.756		2,664
3	3,524		0.658		2,319
4	3,524		0.572		2,016
5	3,524		0.497		1,751
Average	3,524	=	3.353		divided into 11,816

used as a basis for comparison of assets having similar functions but requiring, perhaps, that a larger initial investment be spread over a longer useful service life. An example of this situation is shown in Table 9.5; this asset, costing $15,000, is expected to last for 10 years. All costs are computed in the same way as in the previous example. Converting this 10-year series of costs into the single equivalent annual figure requires discounting the same as before (see Table 9.6).

The resulting equivalent annual cost in this instance is $19,858 divided by 5.019, or $3957. This amount, discounted annually at the 15-percent rate over the 10-year period, has approximately the same net present value as the preceding nonuniform series; the calculations are shown in Table 9.7.

The asset costing $10,000 and having a 5-year life is, therefore, a slightly better choice than the 10-year asset costing $15,000 when all capital-related costs are considered. The higher-priced asset has an equivalent annual cost of $3957, compared with $3524 for the asset with the lower initial cost.

This approach can be employed in a wide variety of engineering economy studies and provides a valid, convenient, and easily understandable

Table 9.5 Annual Costs for an Income-Generating Asset Having $15,000 First Cost and 10-Year Life

| | Account Balance (R) | | | | Annual Charges on Capital | | | |
Year	January 1	December 31	Average Balance	Recovery Charges	Property Taxes (2%)	ROI (15%)	Income Taxes (12.5%)	Total Annual Cost
1	15,000	13,500	14,250	1,500	285	2,137	1,781	5,703
2	13,500	10,800	12,150	2,700	243	1,823	1,519	6,285
3	10,800	8,400	9,600	2,400	192	1,440	1,200	5,232
4	8,400	6,300	7,350	2,100	147	1,103	919	4,269
5	6,300	4,500	5,400	1,800	108	810	675	3,393
6	4,500	3,000	3,750	1,500	75	563	469	2,607
7	3,000	1,800	2,400	1,200	48	360	300	1,908
8	1,800	900	1,350	900	27	203	169	1,299
9	900	300	600	600	12	90	75	777
10	300	0	150	300	3	23	19	345

Table 9.6   Present Value of Annual Costs ($15,000 Asset with 10-Year Life)

Year	Total Annual Cost ($)	×	Discount Factor (15%)	=	Present Value of Annual Cost ($)
1	5,703		0.870		4,962
2	6,285		0.756		4,751
3	5,232		0.658		3,443
4	4,269		0.572		2,442
5	3,393		0.497		1,686
6	2,607		0.432		1,126
7	1,908		0.376		717
8	1,299		0.327		425
9	777		0.284		221
10	345		0.247		85
Average	3,957	=	5.019		divided into 19,858

Table 9.7   Equivalent Annual Costs ($15,000 Asset with 10-Year Life)

Year	Total Annual Cost ($)	×	Discount Factor (15%)	=	Present Value of Annual Cost ($)
1	3,957		0.870		3,443
2	3,957		0.756		2,991
3	3,957		0.658		2,604
4	3,957		0.572		2,263
5	3,957		0.497		1,967
6	3,957		0.432		1,709
7	3,957		0.376		1,488
8	3,957		0.327		1,294
9	3,957		0.284		1,124
10	3,957		0.247		977
Average	3,957	=	5.019		divided into 19,860

comparison between alternatives, especially when they differ only in their original cost and expected service life.

LEASING

Leasing is an important way of financing major assets by means of periodic payments over the asset's useful life. By leasing, the beneficial use of property can be acquired for a specified period of time and for an agreed payment; therefore, the benefits of ownership can be enjoyed with little or no capital outlay.

Leasing substitutes an annual expense for a capital investment. Thus, it offers some of the advantages of a high-capital, low-operating-cost venture without incurring the high initial investment. Leasing is a way to assure equivalent annual costs by a contractual arrangement.

If the property being leased is nondepreciable (such as land), the lease payment represents only interest on the property's economic value. When a depreciable property is leased, the rental payment must cover both principal and interest, so that the asset can be paid off by its owner, with interest, during the leasing period. When a lease covers both depreciable and nondepreciable property (land and a building, for example,) the annual rate is a composite figure, reflecting both elements. The lease-versus-buy alternative is frequently available to a company in the process of acquiring new fixed assets. Properties commonly leased include supermarkets, office buildings, airplanes, automobiles, railroad cars, office copiers, and computers. These properties have widely varying characteristics, yet all of them possess certain qualities that can offer benefits to both the lessor and lessee in a suitable leasing arrangement.

Probably the most common reason for leasing rather than buying a property is to avoid the initial capital expenditure. However, debt financing can accomplish the same objective, spreading out the payments in equal amounts over the asset's useful life, much the same as a leasing arrangement. Another reason for leasing is to allow an organization to use some equipment that it will need only for a short while or that will soon become obsolete. In such instances, there is no option to buy.

The entire rental payment on a lease is tax-deductible; however, debt financing can again produce the same effect. The interest portion of annual debt repayment is deductible, and the principal payments can be offset by annual depreciation or recovery charges.

Leasing a depreciable property, then, does not reduce a firm's capital expenditures; it only rearranges them and places them in a different account. Whether the capital is drawn from capital or operating accounts makes no difference in the rate of return earned by the purchased or leased property, except as affected by the timing of recovery charges and interest payments.

Land and other nondepreciable properties are a completely different situation. Here, outright ownership offers no tax relief, and considerable capital may be tied up nonproductively. By leasing land, the annual payments become tax deductible and the total funds committed are likely to be much lower than if debt financing were employed. Potential opportunities for capital gains through appreciation of land values are of course sacrificed if land is leased rather than owned outright. For this reason, many companies prefer land ownership, even without immediate tax advantage.

Because of income tax considerations, the situation may be such that the landowner cannot justify using the land, and the land user cannot justify owning it. This has led to a number of interesting tax shelters, where land is owned by one group, either a partnership or corporation, and leased to another group, with both groups including the same individuals.

In general, it is more economical for a firm to own depreciable assets and to lease nondepreciable properties. Table 9.8 shows the annual percentage lease rates on assets leased for various periods of time.

In the table, the annual pretax return refers to the interest rate earned by the owner of the property being leased. Thus, if the owner of a building wishes to earn a pretax return of 10.0 percent on his invested capital, he would charge 11.0 percent annually on a 25-year lease, or 13.1 percent annually on a 15-year lease. If a nondepreciable property (e.g., the land on which the building rests) were being leased, the annual rental needed to produce a pretax return of 10.0 percent would, of course, be 10.0 percent.

Typically, supermarkets and similar commercial buildings are leased to yield the property owner a percent or so more than his cost of debt financing. After the lease is up, he will have a building on his hands, and his expectations for the future development of a particular neighborhood will undoubtedly be an important consideration. The duration of a lease de-

Table 9.8   Annual Percentage Lease Rate for Asset Depreciated over Leasing Period

Lessor's Pretax Return	Leasing Period (Years)					
	5	10	15	20	25	30
6	23.7	13.6	10.3	8.7	7.8	7.3
7	24.4	14.2	11.0	9.4	8.6	8.1
8	25.0	14.9	11.7	10.2	9.4	8.9
10	26.4	16.3	13.1	11.7	11.0	10.6
12	27.7	17.7	14.7	13.4	12.8	12.4
15	29.8	19.9	17.1	16.0	15.5	15.2
20	33.4	23.9	21.4	20.5	20.2	20.1
25	37.2	28.0	25.9	25.3	25.1	25.0

pends on the characteristics of the property. Equipment subject to a high rate of obsolescence leases for a short time at a high rate; electronic data processing equipment, for example, is usually priced to pay out in 2–4 years on short-term leases, since the market for outdated computers is not strong. Similarly, supermarkets seldom remain at a given location for more than 15 years, and abandoned buildings must often be demolished to restore the land value.

Making a sound decision regarding the terms of a leasing agreement, or even reaching a lease-versus-buy decision, is particularly troublesome when there are intangible considerations involved. The mathematics of the transaction are straightforward enough, but it is difficult to assign realistic values to these intangible factors.

SUMMARY

Fixed and variable costs, common to all businesses and industries in widely varying proportions, significantly affect a company's economic position and competitive strategy. Some businesses are characterized by high fixed costs and low operating costs, whereas others take the opposite position. In general, a $5 investment can reduce direct labor costs by about $1 annually, thus allowing management considerable flexibility in establishing a firm's overall cost structure. In identifying a desirable fixed-variable cost relationship for either a firm as a whole or a single new venture, or in comparing alternative projects having different cost characteristics, it is important to translate these two types of costs into a single, equivalent annual amount. For this purpose, the amortization concept is useful. An initial capital expenditure can be converted to a series of annual payments by applying appropriate rates for depreciation, interest, *ad valorem* taxes, G & A expenses, ROI, and income taxes. Then the annual amounts can be converted to an equivalent annual basis by applying an appropriate discount rate to each annual payment, summing the present values of the future annual amounts, and dividing by the sum of the annual discount rates. As an alternative to outright ownership of capital assets, leasing offers a means of contractually obtaining equivalent annual costs by substituting a uniform yearly payment for an initial capital outlay.

10

Forecasts and Forecasting

Business more than any other occupation is a continual dealing with the future; it is a continual calculation, an instinctive exercise in foresight.
Henry R. Luce

The term *forecast* has several dictionary definitions:

to plan ahead;
to foresee; to calculate beforehand;
a prophecy or estimate of a future happening or condition.

Forecasting, then, means that conclusions regarding future events require some sort of predictive model or technique that can use today's data. If the facts are correct and the technique is valid, the conclusions will be accurate.

For those involved in the economic evaluation of engineering projects, forecasts of three broad types of "future conditions" are particularly important: economic, technological, and business. *Economic conditions* refer to the nation's aggregate economy, without particular reference to any specific industry. The gross national product (GNP) is probably the most important single indicator of general economic conditions. *Technological conditions* encompass a variety of factors affecting the development of new products, processes, and markets. *Business conditions* take into account the specific factors of price, cost, and volume, which ultimately determine the profitability of a given industry, company, or venture.

MANY SOURCES OF DATA—SOME FREE

The first step in forecasting is to collect the appropriate data. The most common sources of economic data include government agencies, trade associations, and business publications. When no published data are available, special tabulations or surveys may be warranted.

Much useful economic data can be found in readily available publications. The following list of information sources is not meant to be all-inclusive, but a familiarity with the references included in this list will prove extremely useful to anyone concerned with statistics that specify economic conditions.

Directories

Statistics Sources. Gale Research, Detroit, MI.
Executive's Guide to Information Sources. Gale Research, Detroit, MI.
Business Trends and Forecasting Information Services. Gale Research, Detroit, MI.
Encyclopedia of Associations. Gale Research, Detroit, MI.
Thomas Register of American Manufacturers. Thomas, NY.
Poor's Directory of Executives. Standard & Poor, N.Y.
Million Dollar Directory. Dun & Bradstreet, NY.

General economic data

Statistical Abstract of the United States. U.S. Government Printing Office, Washington, D.C. (annual).
Economic Almanac. National Industrial Conference Board, NY (annual).

Survey of Current Business. U.S. Office of Business Economics, Washington, D.C. (monthly).

National Economic Projections. National Planning Association, Washington, D.C.

Economic Indicators. U.S. President's Council of Economic Advisors, Washington, D.C. (monthly).

Handbook of Basic Economic Statistics. Economic Statistics Bureau, Washington, D.C. (monthly).

World Almanac and Book of Facts. Newspaper Enterprise Association, New York, NY (annual).

Industry and product data

U.S. Industrial Outlook. U.S. Bureau of Defense Services Administration, Washington, D.C. (annual).

Predicasts. Predicasts, Cleveland (quarterly).

Census of Business. U.S. Bureau of the Census, Washington, D.C.

Census of Manufactures. U.S. Bureau of the Census, Washington, D.C.

Census of the Mineral Industries. U.S. Bureau of the Census, Washington, D.C.

Census of Population. U.S. Bureau of the Census, Washington, D.C.

Minerals Yearbook. U.S. Bureau of Mines, Washington, D.C. (annual).

Construction Review. U.S. Bureau of Defense Services, Administration, Washington, D.C. (monthly).

Area information

Commercial Atlas and Marketing Guide. Rand-McNally, Chicago.

Editor and Publisher Market Guide. Editor and Publisher, NY (annual).

Survey of Buying Power. Sales Management Magazine, NY (annual).

Municipal Yearbook. International City Manager Association, Chicago (annual).

Financial data

Quarterly Financial Report for Manufacturing Corporations. U.S. Federal Trade Commission and U.S. Securities and Exchange Commission, Washington, D.C. (quarterly).

Leading U.S. Corporations. News Front Magazine, NY.

Moody's Handbook of Common Stocks. Moody's Investors Service, NY (quarterly).

Stock Guide. Standard & Poor, NY (monthly).

Value Line Investment Survey. Arnold Bernhardt, NY (weekly).

General business and financial periodicals

Wall Street Journal. Dow Jones, NY (daily).
Business Week. McGraw-Hill, NY (weekly).
Forbes. Forbes, NY (bimonthly).
Fortune. Time, NY (monthly).
Dun's Review. Dun & Bradstreet, NY (monthly).

Obviously, not all these references will be pertinent to the engineer's interest, and other specialized publications may offer more direct benefits. But anyone involved in economic analyses should be aware of anticipated changes in the economic environment.

BASIC FORECASTING TECHNIQUES

Forecasting is said to be "sometimes a science, sometimes an art, and most often a little of both." Essentially an economist's tool, forecasting is a necessary part of planning most engineering projects.

The accuracy of economic forecasts can often mean the difference between business success and failure. Management decisions in the areas of production, financing, marketing, inventory planning, and capital investment depend to a large extent upon the information developed in economic forecasts.

Several assumptions are helpful in approaching any business forecast:

1. The magnitude of, and the relationships between, most economic factors change slowly.
2. What happens in the future is strongly influenced by what is going on at present.
3. Past experience offers the best guide to these cause-and-effect relationships.

In predicting the future, then, it is necessary to "know the past and understand the present." A forecast of economic activity in any business or industry involves an analysis of past trends, extrapolation of these trends into the future, and adjustment, refinement, or revision of projected trends to allow for deviations.

The specific techniques employed in economic forecasting depend largely upon how far into the future the forecast extends. For extremely short-range predictions—perhaps covering only a week or a month—it may be safe to assume that the immediate future will be very much like the present, or at least that recent trends will continue for the period covered by the forecast. This approach proves adequate most of the time, with significant errors occurring only when there is a major reversal of trends.

A second method, frequently used for intermediate-term forecasts, employs "leading" indicators to predict coming changes. Next year's construction activity, for example, will be indicated by this year's contract awards. Some of the most important indicators used by economists include interest rates, contract awards, stock prices, manufacturers' new orders, freight car loadings, indexes of industrial production, corporate profits, and business failures.

A similar procedure can be used effectively with "concurrent" indicators, providing that they can be projected with some degree of certainty. Construction volume, for example, can be correlated closely with the GNP. If a good estimate of next year's GNP is available, the total contract construction volume can be estimated within 1 or 2 percent. When the construction-GNP relationship is expressed mathematically, it becomes in effect another type of forecasting technique.

This third type of economic forecast uses an econometric model—a mathematical-statistical description of an economy, a business, or an industry. This approach is especially useful in long-range forecasts, when there are many unknown factors involved. By incorporating these factors, in equations, into an electronic spread sheet on a microcomputer, they can be varied and the effects of change on the ultimate outcome can be calculated. In this way, many different assumptions can be used and their results examined; as a result, the most likely outcomes can be predicted, as well as the range of possible outcomes.

TREND AND REGRESSION ANALYSES ILLUSTRATE A PRINCIPLE

Business economists employ many techniques in making their forecasts, with widely varying levels of precision and sophistication.

There is scarcely a limit to how complicated a problem can be made, how far back data can be collected, or how much data can be included. In many instances, however, the *principle of historical continuity* yields as reliable a prediction as the most sophisticated econometric models.

The principle of historical continuity

When predicting future prices or weather, unless there is some specific reason to believe otherwise, the safest bet is usually to assume that tomorrow will be pretty much the same as today. That assumption expresses the principle of historical continuity. In more mundane terms, it is a direct outgrowth of the ancient "lost horse" technique, which states that, to find a lost horse, go where it was last seen and start looking in whatever direction it was heading. In practice, patterns and trends established in the past often prevail with sufficient frequency to warrant using them as a basis for predicting the future. Extending the already established trend into the future involves *extrapolation*. Identifying the appropriate trend, though, requires some careful thought.

To lend the quality of statistical objectivity to predictions made through extrapolation, a regression line can be fitted to the available data and then extended. The best way to assure a valid fit of the regression line is to apply the principle of least squares.

The method of least squares can be used graphically

In fitting a straight line to a series of points, there is bound to be some scattering of the points about the line. To find a regression line around which the degree of scattering is minimum, the *method of least squares* is commonly used.

A regression line selected according to the principle of least squares is, by definition, the line that results in the smallest possible sum of the squared vertical deviations of the points from the line. An equation for this line can be developed mathematically. However, if only an indication of the magnitude and direction of a trend is desired, a quick graphical solution is usually adequate.

Figure 10.1 illustrates the graphical method of solution. There are seven points having equal horizontal spacing, on which a least-squares regression line is to be fit. Starting from the point on the extreme left, draw a straight line toward the second point, extending this line just two-thirds of the horizontal distance between the two points. Continuing from this point, aim toward the third point and draw a line in that direction for the same horizontal distance—that is, two-thirds of the horizontal distance between each pair of adjacent points. Repeat this procedure point by point until the "target" point is the last one, at the extreme right. The final line segment

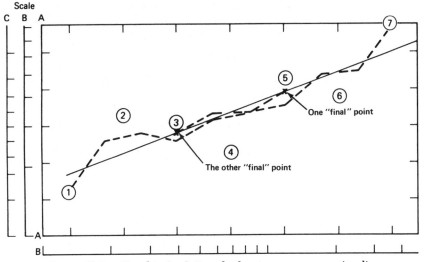

Figure 10.1. *Graphical solution for least-squares regression line.*

drawn will end exactly two-thirds of the horizontal distance from the extreme left-hand to the extreme right-hand point. This final point lies on the
least-squares regression line.

To locate the second point needed to define the regression line, start with
the point on the far right and work back toward the left, point by point,
each time moving a distance equal to two-thirds of the horizontal distance
between adjacent points.

The dotted lines in the figure show the course of the line developed from
left to right, ending in an x just below the fifth point; and from right to left,
ending at the x just below the third point. A straight line passing through
each x represents the least-squares regression line—or the unique straight
line that minimizes the sum of the squares of the vertical distances from
each point to the regression line. The sum of the squared deviations is
known statistically as the variance; its square root is known as the standard
deviation.

Calibration of the scales makes no difference, so long as the points are
uniformly spaced and measurements are made at a constant horizontal increment. Vertically, for example, scale A is arithmetic; scale B is logarithmic; and scale C represents a normal probability distribution.

On the horizontal scale, A is arithmetic and B is logarithmic. However, for the points to be horizontally equidistant on the logarithmic scale, they must be plotted at increments of 2^N—at units of 1, 2, 4, 8, . . . , for example. An arithmetic scale is far more convenient.

TIME SERIES ANALYSIS CAN PORTRAY INDUSTRY CHARACTERISTICS

Time series analysis is one of the most useful applications of this graphical technique in economic analysis. The horizontal scale is ordinarily arithmetic, measuring years; the vertical scale may be either arithmetic or logarithmic, measuring dollars, ratios, or index numbers and indicating either absolute changes (on an arithmetic scale) or rates of change (on a logarithmic scale).

When plotting numerical values by years, how closely the individual points fit the least-square regression line is a direct indication of stability, whereas the slope of the line measures the trend. Most mature public utilities' earnings, for example, plot nearly as a straight line, indicating low variance and high stability and thus offering a relatively high level of confidence in predicting future earnings.

Firms in less stable industries, especially those subject to rapid technological change, intense competition, or highly erratic market demands, are apt to show much greater variances, and projections of their future earnings are, therefore, subject to a much higher degree of uncertainty.

EXTRAPOLATION

The best guide to an individual's future behavior is often indicated by his past behavior in similar situations. Such is also the case with economic data; the best estimates of the future can often be made by a careful study of the past. Unless there is some reason to believe otherwise, whatever trends have taken place in the past can generally be expected to continue into the future. This technique of economic forecasting, in which established trends are assumed to continue, is called extrapolation.

Extrapolation refers simply to the extension of a series of numbers beyond its original range by adding other numbers, at either end of the series, which fit the established pattern of the original series. In other words, the values of points lying beyond a known interval are estimated solely on the basis of known values lying within the interval, without regard to any underlying causes.

Extrapolation is the simplest, yet one of the most effective and useful, forecasting techniques. It is most commonly applied to time series data, but can also be used in other types of correlation analyses. Since the basic assumption in forecasting by extrapolation is that the future is a direct reflection of past trends, the basic problem is to identify the appropriate growth trend indicated by historical data. This is done in several steps:

1. Plot the pertinent data on graph paper.
2. Try different curves to see which fit the data.
3. Select the curve that fits best.
4. Fit the appropriate trend line to past data, using some accepted statistical technique.
5. Extend the trend line into the future.
6. Read the expected future values off this extended trend line.
7. Adjust these future values according to any other available knowledge.

Numerous curves can apply to various types of data; however, certain types are especially typical of certain types of economic data. Only a few of the most commonly encountered curves are discussed here.

TRENDS AND STANDARD CURVES

Generally, the most useful trend curves encountered in economic forecasting are the arithmetic, the semilogarithmic, the logarithmic, and the S shape.

These curves are illustrated in Figures 10.2–10.5, and are plotted on arithmetic scales to show their general shapes. They plot as straight lines on different types of graph paper.

THE ARITHMETIC TREND.   The arithmetic relationship shown in Figure 10.2 plots as a straight line on conventional graph paper. It represents, in a time series, a constant numerical increase per unit of time. The arithmetic curve has the general form of the linear equation, $y = a + bx$.

THE SEMILOGARITHMIC TREND.   The semilogarithmic trend (Figure 10.3) depicts a constant *rate* of change (as opposed to a constant numerical change in the arithmetic trend). Thus, the change occurring during each time period is a fixed percentage of the cumulative total at the end of the

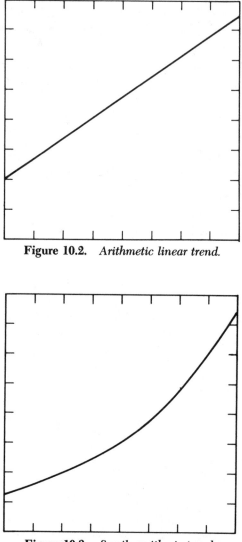

Figure 10.2. *Arithmetic linear trend.*

Figure 10.3. *Semilogarithmic trend.*

preceding time period. The semilog trend is typical of population growth and many other types of economic data, such as per capita consumption of utility services. Values following a semilog trend plot as a straight line on semilog graph paper, which is the most convenient way to extrapolate and present the data. This curve follows the form $y = ab^x$.

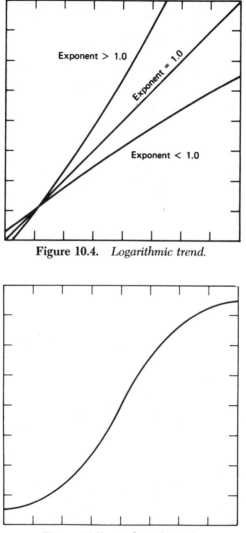

Figure 10.4. *Logarithmic trend.*

Figure 10.5. *S-shaped trend.*

THE LOGARITHMIC TREND.   The logarithmic trend takes an exponential form, as illustrated in Figure 10.4. If the exponent is greater than 1, it will slope upward; if the exponent is less than 1, the curve will slope downward; and if the exponent is exactly 1, the trend will, in effect, be linear and will plot as a straight line either on log-log or arithmetic graph paper. Cost-capacity data typically follow the exponential form, plotting as a straight line on log-log paper and having exponents less than 1. Doubling a facility's capacity, for example, may increase the cost by only half, representing a logarithmic trend having an exponent of about 0.6. Mathematically, this curve can be defined as $y = ax^b$.

THE S TREND.   The S-shaped curve shown in Figure 10.5 plots cumulatively a "normal" or binomial frequency distribution of data. This type of curve is especially valuable in forecasting future sales of a specific product, since the life-cycles of most products generally follow this shape. They typically start slow, pick up momentum as they gain market acceptance, then taper off gradually as the market becomes saturated and newer products begin replacing them. A symmetrical S curve plots as a straight line on probability graph paper. To determine where a specific product stands on its growth curve, an estimate must first be made of its total potential cumulative sales; its cumulative sales history to date (expressed as a percentage of the estimated potential) can then be plotted on probability paper. If the resulting plot is a straight line, the estimate was good; otherwise, a new estimate of sales potential should be made and the entire procedure repeated until the plotted points come out on a straight line. Extending the straight line gives both the number of years until the potential is reached and the annual sales increments from which the sales for any given year can be calculated. Other common types of S-shaped curves include the logistic and Gompertz curves, and many theoretical cumulative frequency distributions plot in the S shape.

TECHNOLOGICAL FORECASTING ANTICIPATES INTERDEPENDENT CHANGES

Intelligent planning of engineering projects requires numerous assumptions about the future of technology, so that the impact of future changes on present designs can be anticipated. Such asusmptions require prediction of what is by definition essentially unpredictable: innovation and technological change.

The only certainty about the future is that it will be different from the present, just as certainly as the present is different from the past. Determining just how much variation can be expected, and defining the ways in which the changes will be manifested, are the primary objectives of technological forecasting. Whether the engineer's interests are in water supply and wastewater disposal, electric power generation, transportation systems, environmental pollution, or other fields, technological forecasts can play an important part in decision making.

Technological change is a result of economic, social, and political conditions that largely determine the extent and direction of technological effort. Consequently, the first step in making a technological forecast is to analyze what had to have taken place for technology to arrive at its present status. The future can then be predicted, based on where society in general is likely to be headed. A concern with environmental quality, for example, is certain to have significant effects on technology related to wastewater treatments, solid waste disposal, air pollution control, and noise abatement programs.

The second step in the technological forecast is to predict both the qualitative and quantitative changes in the technology under consideration. Using air pollution as an example, the quantity of particulate matter discharged into the atmosphere by a particular industry in the year 2000 will be a function of at least three independent factors: (1) changes in the amount of pollutants produced, brought about by changes in production levels and processes; (2) changes in the number or proportion of plants in that industry that have installed particulate control devices; (3) the collection efficiency of the control systems in operation. Production levels depend chiefly upon economic considerations; pollution control practices are most influenced by legal restraints in the form of politically established emission control standards; and production processes and equipment efficiency are engineering functions.

The third step, then, is to reconcile the results of the second-step forecast with the requirements defined in the first step. In the preceding example, if emisson levels are still higher than society will accept, the original forecasts must be modified. Either technology must conform to society's demands, or society must adapt itself to technology. The forecaster must somehow end with a set of forecasts—economic, sociological, and technological—that are compatible.

TECHNOLOGICAL EFFICIENCY ALSO FOLLOWS AN S-TREND

The efficiency of almost everything increases with time (*time*, not age). A 1980 generating plant is more efficient than a 1940 plant was when the latter was new; and a generating plant built in the year 2000 will be more efficient than the plants installed in 1980.

The following univeral forecasting technique for technological efficiency is offered for use in situations in which no better data are available. The curve in Figure 10.6 represents the improvement in technological efficiency that is achieved with time. The vertical scale measures efficiency, and the horizontal scale represents the time interval required to achieve, in practice, any desired increase in efficiency.

The curve indicates, for example, that if something is currently operating at 50-percent efficiency, its efficiency should rise to 62 percent in 10 years. Things presently at a 75-percent efficiency level are expected to increase to 84 percent in 10 years; but if the efficiency is already 95 percent, the expected 10-year improvement will raise it only to 98.5 percent. Nothing is ever 100 percent efficient.

This curve does not necessarily imply that a specific product will be

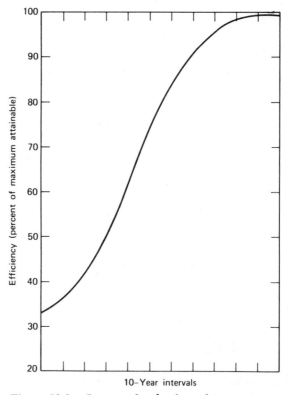

Figure 10.6. *Pattern of technological improvement.*

more efficient in the future than it is now. On the contrary, the curve is based on the assumption that more efficient ways will be found of performing the same function. Thus, the curve does not mean that the efficiency of mechanical dust collectors will increase, say, from 80 to 88 percent during the next decade. It does, however, indicate that if the average efficiency of particulate control systems currently being installed on industrial plants is 80 percent, the efficiency of systems installed 10 years from now will average 88 percent. The increase in system efficiency may partially result from increased cyclone efficiency, but will probably be more strongly influenced by a shift toward electrostatic precipitators, fabric filters, and other types of more efficient equipment.

Similarly, near-term increases in the efficiency of wastewater treatment plants will be a result of more widespread use of presently available technology, rather than any innovative treatment concepts. To raise the efficiency of sewage treatment, activated sludge systems will prevail over trickling filters to an increasing degree, new treatment processes will be developed, and provisions for tertiary treatment will become more prevalent.

Technological change does not necessarily guarantee technological improvement, and what technology is capable of doing is not necessarily what technology does. Nevertheless, the technological forecast can provide the engineer with much information useful in long-range planning and can help him make intelligent decisions in view of an always uncertain future.

FORECASTING BUSINESS SUCCESS RELIES ON THREE COMPONENTS

Most new business ventures fail outright; many others never yield the returns that were expected for them. Yet the great majority of these business failures can be attributed to circumstances or conditions that were known prior to undertaking the project, or to factors that could easily have been identified well in advance of making any major financial commitments. For some reason, whether enthusiasm or ignorance, management simply failed to recognize the importance or existence of these factors.

To be financially successful, a new product or new service venture must have three broad bases of support:

1. A superior product.
2. A capable company.
3. A favorable environment.

The whole problem of forecasting business success in a particular venture, whether it involves a new product or a new service, can be visualized conceptually as a three-legged milking stool—the product, the company, and the environment representing the legs supporting the overall venture (see Figure 10.7). A single short or weak leg can cause the entire venture to topple, or at least to become uncomfortably shaky.

The specific problems encountered in different types of business activity vary widely and the relative importance of factors are not the same in each case; however, the primary determinants of success are very much alike: *what* is being sold, *who* is attempting to sell it, and *where* it is sold. In other words, the *product*, the *company*, and the *market* together determine how successful the venture will be.

The *product* or service must be objectively evaluated in terms of its competitive performance, salability, and defensibility. Each of these categories is further divided into several individual factors (such as effectiveness, reliability, simplicity, convenience, appearance, and many others) to determine how favorably the proposed product compares with the products against which it must compete in the marketplace.

The *company's* capabilities must include three broad areas: marketing, technology, and production. Marketing is probably the most important of the three, and, each can be broken down into many distinct factors which can be evaluated individually.

The *environment* refers to external considerations such as the market, competition, suppliers, and government. The company must react to en-

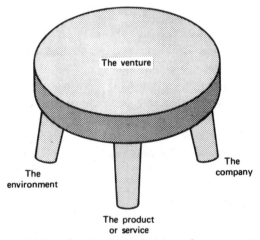

Figure 10.7. *The elements of success for new ventures.*

vironmental conditions, but can seldom control them; it must sell its products in the existing environment, subject to numerous conditions and constraints imposed by competition, legislation, and the general economy. As before, the pertinent factors can be readily identified and evaluated for a specific venture.

Finally, the overall *venture's* success is measured in terms of its investment returns, strategic benefits, and other outcomes determined by a combination of the product, company, and environmental factors.

By putting together all the factors relating to product success, it is possible to predict, in advance and with fair reliability, the probability of the product's ultimate success in the marketplace. Table 10.1 gives a rough indication of the relative probability of success for new product ventures, based only on whether the proposed new product is better than, equal to, or not as good as the competition, with respect to the three broad rating factors—the product, the company, and the environment.

The assumption is made in this table that all three factors are of equal importance. A venture rating high in all three categories is assigned a relative probability of success of 100, and all other combinations are related to this index value.

The techniques by which several hundred individual factors can be incorporated in the venture evaluation model can become quite sophisticated and far more useful and meaningful than the ultrasimple method summarized in the table. For a more complete discussion of this whole approach to venture evaluation, the reader is referred to *Strategic Analysis for Venture Evaluation,* by W. R. Park and J. B. Maillie (Van Nostrand-Reinhold, New York, 1982).

Even a strong company cannot successfully sustain a market for a poor product, nor can a company with inadequate capabilities in marketing be successful even with a superior product. It is possible for a first-rate company to introduce a product that is far superior to all other available products and then find that there is no consumer demand for the product, regardless of its superior attributes. Only if a strong, capable company offers a superior product in a well-defined market in which a strong demand exists, is there a good chance of success.

There are of course exceptions. Whenever a new company is formed, it faces a handicap—similar to a pinch hitter going into a baseball game with a one-strike count. A new company with a good product in an uncertain market has two strikes against it at the beginning, as would an established company with an inferior product in a weak market, or a company with limited marketing capabilities and an inferior product in a strong market.

Table 10.1   The Probability of Success for New Product Ventures

Rating Factors[a]			Relative
The Product	The Company	The Environment	Probability of Venture Success
+	+	+	100
+	+	=	60
+	+	−	25
+	=	+	60
+	=	=	40
+	=	−	15
+	−	+	25
+	−	=	15
+	−	−	5
=	+	+	60
=	+	=	40
=	+	−	15
=	=	+	40
=	=	=	25
=	=	−	10
=	−	+	15
=	−	=	10
=	−	−	3
−	+	+	25
−	+	=	15
−	+	−	5
−	=	+	15
−	=	=	10
−	=	−	3
−	−	+	5
−	−	=	3
−	−	−	1

[a] + Represents good, or better than competition; = represents average, or comparable to competition; − represents poor, or inferior to competition.

Still, there have been instances of new companies offering new products in untested markets and being successful, especially in low-priced "fad"-type consumer items. There are, however, many more examples of such ventures that have failed miserably.

It is possible to take all the hundreds of individual factors affecting the success of a new product, service, or venture, put them together in a structured, well-ordered model, and draw some meaningful conclusions from them. Adjustments can be made for the type of product or industry; uncertainties in judging certain factors can be considered; and operational differences in companies and unusual conditions can be anticipated and incorporated in the model. Employing an objective, analytical approach such as this to venture evaluation has, in practice, repeatedly shown several important advantages:

1. It provides a convenient, economical, and versatile method for quickly and objectively screening large numbers of proposed new ventures.
2. It compares the relative merits of proposed ventures on a consistent and well-defined basis, both with each other and with existing product lines or services.
3. It directs management attention to all factors affecting venture success, thus preventing enthusiasm for a few favorable points from carrying a proposal through.
4. It effectively concentrates and utilizes the collective observation, experience, and judgment of company management.
5. It identifies and measures present and potential strengths and weaknesses.
6. It reduces the chances of failure by pinpointing the potential causes of failure for a specific venture at the beginning.
7. It maintains continuity and consistency in a firm's overall venture evaluation program.

THE BUSINESS PLAN IS THE BUSINESSPERSON'S MAP

Every business venture, new or old, whether leading its industry or trying to catch up with it, needs a plan.

The business plan is a strategic map that shows where a business intends to go, and defines the route by which it expects to get there.

This plan is essential for every business; however, its purpose may vary. For a new venture seeking outside financial resources for seed money or start-up capital, the main purpose of the business plan may be to present tangible evidence of the venture's economic desirability as an investment to prospective sponsors. For an in-house new venture or diversification opportunity, the business plan may be intended to convince the firm's management that the project is worth being pursued and funded. For an established business, the plan's main benefit may be simply to insure that the management team is thinking along similar lines and working toward specific, well-defined goals. For a small business—the kind that is often drifting along without any particular goal other than continued survival—the business plan forces management to think about competitive conditions, marketing opportunities, and potential problem areas.

Although business success may begin with, and is often attributed to, some extraordinarily brilliant concept, it can actually be achieved only by giving considerable attention to, and carrying out, thousands of piddling little details. The business plan should contain at least the most important of these details, preferably in 40 pages or less. As mentioned, the content can be expected to vary, depending on the primary purpose of the document.

The content of a somewhat typical business plan is described in the following sections. This particular plan is for a new venture to be built around a proprietary new product. It is directed toward venture capital firms and other prospective investors. Its purpose is to raise funds for the research and development activities needed before the product can be brought to market.

1.   SUMMARY OF PROPOSED R & D PROGRAM.   The five-year R & D program is broken into four distinct phases. This section briefly describes the activities, time, and cost associated with each phase.

2.   PROPOSAL FOR FINANCIAL PARTICIPATION.   Here, the maximum financial commitment, the probable cost, and the equity position associated with an investment in each phase is defined. Cash flow projections show the payout time, payout ratio, and internal rate of return to prospective investors.

3.   ABSTRACT OF BUSINESS PLAN.   This section actually comprises an "executive summary" of the entire document. In a single page, it describes the proprietary process, defines its market, tells what progress has already been made and how much money and time is needed to carry the project until it can earn its own way, and mentions the most important strategic considerations confronting the venture.

4.   LEGAL AND PATENT INFORMATION.   A key factor in the success of any proprietary product is its legal defensibility, whether by patents or other means. This section, therefore, will be of great interest to the prospective

investor. In this case, copies of two documents are included: (1) a letter of opinion from the firm's patent attorney, discussing the current status of the various patents and patent applications; and (2) a "Notice of Allowance" from the U.S. Patent Office, showing that the patent applications have been examined and found allowable for issuance of patents. If desired, a copy of the actual patents issued could be included either in an appendix or as a separate document.

5.   BACKGROUND INFORMATION.   This section provides an introduction to the body of the business plan, including a brief history of the firm, how the proposed new product evolved, and a description of its salable benefits, its value and importance, and its applications.

6.   MARKET ANALYSIS.   Here, the market potential for the new product is evaluated. The analysis should comprise data from credible sources on total industry volumes and trends, probably including both domestic and foreign markets. Ten-year sales projections are made by applying appropriate share-of-market assumptions, and a marketing plan is then presented, defining how the firm expects to achieve its projections. If market penetration is based on the firm's having marketing advantages in terms of cost, price, location, resources, or some other measurable item, then these advantages should be described.

7.   STRATEGIC ANALYSIS.   The strategic analysis is divided into four parts: (1) the product, (2) the company, (3) the environment, and (4) the venture itself. Product factors to be considered include its performance, salability, and defensibility. Company factors include the firm's capabilities and resources for marketing, technology, and production. The environment encompasses the market, the competition, suppliers, and government. Venture factors cover the venture's support, investment, and overall strategy. Each factor in the strategic analysis should be considered in terms of the competition in general, and rated as being better than, worse than, or equal to competition. This section will identify the venture's major strengths and weaknesses.

8.   PROJECT BUDGET.   In this section, the project budget may be presented in several different ways, in different levels of detail. First, the budget is shown in summary form, with yearly totals broken down between operations and equipment, and with appropriate allowances for working capital, financing fees and expenses, and contingencies. A detailed breakdown of major equipment or plant costs is presented next. Then, the annual budgets are broken down into quarterly (although some prefer monthly)

amounts over the duration of the project. Finally, the annual compensation for the principals involved in the project is presented.

9.   CORPORATE ORGANIZATION.   An organization chart, showing the identity and functions of the principals involved in the project—both directly and as advisors—is desirable. The manner in which the project is organized may have a direct impact on how it might best be financed.

10.   BIOGRAPHICAL INFORMATION.   Prospective investors may lack the technical background necessary to thoroughly understand how a new technology functions, but they can understand a great deal about a project's chances for success simply by knowing as much as possible about the people involved in the project. Many venture capital firms will, in fact, devote more effort to investigating the people than to the product itself.

11.   APPENDIX.   If additional technical explanation is needed, an appendix is the place to put it. Often, the people studying the business aspects of the proposal will be different from those concerned with its technology. Whatever other relevant information is considered necessary can also be included here.

The business plan described here is just one type of many. Business plans intended for other purposes are likely to include many of the same items, along with a number of considerations not covered here. This emphasizes that a business plan is necessarily unique to the specific business for which it is being developed, so it makes sense to pick out and include only the items that are appropriate for the specific organization and purpose.

Businesses never plan to fail, but they often fail to plan. Even a bad plan is better than no plan at all. Nevertheless, the pressure to cope with immediate problems—"putting out fires"—often makes it difficult to spend the necessary time carrying out the planning that might keep those fires from occurring in the future. So much time is spent on matters that are urgent, there is little left for matters that are really important.

The business plan will help to identify those important matters.

SUMMARY

Forecasting involves making predictions of future events or conditions based on present knowledge. It is especially important and useful to the engineer working in economic, technological, and business areas, and is accomplished by applying an appropriate forecasting technique to available, reliable data. The forecasting technique generally involves analysis of past trends, extrapolation of these trends into the future, and adjustment or refinement of the resulting projections. Much useful economic data can be obtained from readily available published sources. Once the necessary data have been accumulated, basic statistical techniques can be applied for trend and regression analysis, time series analysis, and extrapolation. Technological forecasting, an important part of many engineering proposals, requires analysis of the impact of economic, social, and political considerations as they influence technological development and change. Success in a new business venture is dependent on three broad bases of support: (1) a superior product or service, (2) a capable company, and (3) a favorable environment in which to operate. Being deficient in any of these three areas substantially reduces the chances for success, and a careful evaluation of all pertinent factors is therefore essential in considering the potential attractiveness of a new venture. The business plan incorporates all of these considerations in a written document that provides an overall strategic map of the business.

11

Probability, Risk, and Uncertainty

It is a truth very certain that, when it is not in our power to determine what is true, we ought to follow what is most probable.

Descartes

Most management decisions are based on a combination of experience and anticipation—sometimes dignified with estimates of probabilities. In most cases, the application of statistical techniques and probability theory can do much to help define uncertain events in objective, measurable terms.

The usefulness of statistics as a tool for engineering-economic analysis depends to a large extent upon the availability of suitable data, since the inferences to be drawn from statistical analysis can be no better than the raw data used as a basis for the study. Typical data suitable for statistical treatment include information on sales, estimated and actual costs, and profits.

Data accumulated for economic analysis and for eventual use in business planning are more easily interpreted if they are arranged logically. Usually, the arrangement includes compilation of frequency distributions and construction of a frequency distribution curve.

THE FREQUENCY DISTRIBUTION

A frequency distribution is a grouping of statistical data after the data have been sorted and arranged in some order, such as from the smallest to the largest. The frequency distribution shows how often a given item of data occurs within each group. The frequency distribution is particularly valuable in analyzing relationships involving large quantities of data that might otherwise be difficult to interpret. Given the frequency distribution, both the average values of a mass of data and the entire range of values can be clearly defined.

In constructing a frequency distribution curve for a large mass of data, raw data are grouped into their appropriate classifications by dividing the overall range of values covered by the data into convenient-sized groups and then tallying the data falling within each group. Table 11.1 illustrates

Table 11.1   Accuracy of Engineer's Estimates

Range of Ratios—Actual Cost to Estimated Cost	Projects	
	Ratios in This Range (%)	Ratios above This Range (%)
0.75–0.80	3.0	97.0
0.80–0.85	15.2	81.8
0.85–0.90	19.7	62.1
0.90–0.95	10.6	51.5[a]
0.95–1.00	15.2	36.3
1.00–1.05	15.2	21.1
1.05–1.10	10.6	10.5
1.10–1.15	4.5	6.0
1.15–1.20	3.0	3.0
1.20–1.25	3.0	0.0
	100.0	

[a] Sample calculation (from center column): 3.0 + 3.0 + 4.5 + 10.6 + 15.2 + 15.2 = 51.5

some grouped data used in analyzing the accuracy of engineers' estimates. The raw material for Table 11.1 consisted of the ratios of actual cost to engineers' estimated cost for a large number of jobs. The ratios were grouped into 10 narrower ranges, each representing an interval of 5 percent of the estimated cost. The mean of the ungrouped data was 0.96, although the individual jobs cost from 25 percent more to 25 percent less than estimated. More than half the jobs fell within ± 10 percent of their estimated costs, and 30.4 percent were within 5 percent of the amount estimated.

Figure 11.1 shows the frequency distribution in graphical form as it is plotted from the data in Table 11.1. This figure is generally typical of frequency distribution curves in that the curve rises from zero to a maximum somewhere around its average value (in this case, the highest frequency of occurrence is in the 0.85–0.90 interval), tapers off, and finally reaches zero again. If the curve were perfectly symmetrical, it would usually be referred to as a normal curve representing a normal distribution, and the highest point on the curve would correspond to the arithmetic average of all the data. In statistical terms, the distribution of Figure 11.1 is "positively

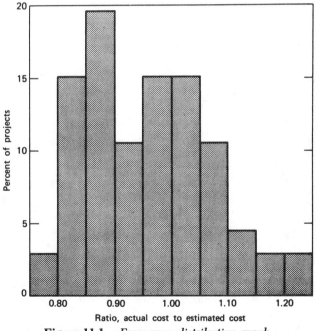

Figure 11.1. *Frequency distribution graph.*

skewed," signifying that the distribution tails off toward the right at a slower rate than toward the left, and meaning that the range of values on the high side of the average value is greater than the range of values on the low side.

THE CUMULATIVE FREQUENCY DISTRIBUTION FOR SETTING CONTINGENCIES

The right-hand column in Table 11.1 represents the total number of jobs having actual/estimated cost ratios equal to or greater than the range given in the left-hand column. This total is found by adding together the total number of estimates falling within each interval, starting from the end and working backward.

The cumulative total number of jobs falling above each range of values is plotted in Figure 11.2. The frequency distribution data, plotted cumulatively, start at 100 percent of the total number of jobs included in the

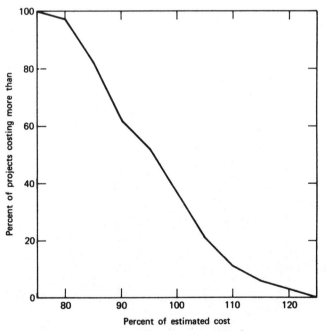

Figure 11.2. *Cumulative frequency distribution curve.*

analysis. From this point, the curve decreases rapidly until it reaches approximately the middle, at which time the rate of decline decreases and the curve tapers off gradually until reaching zero.

The cumulative frequency distribution curve can be used in predicting the probability of an event's occurrence, assuming that historical data are indicative of the future. For example, experience has shown in this case that 36.3 percent of the jobs cost more than the engineers' estimates. By adding a 5-percent "contingency" allowance to the estimated cost, the percentage of jobs costing more than estimated drops to 21.1 percent; a 10-percent contingency allowance covers all but 10.5 percent of the jobs; and a 15-percent allowance leaves only 6.0 percent of the jobs exceeding the estimate. Although estimating accuracy is not improved by inclusion of this contingency allowance, the probability of exceeding the estimate is reduced substantially, and the practice may be worthwhile from the standpoint of capital budgeting.

PROBABILITY AND EXPECTATION

Probability refers to the frequency of occurrence of an event, measured as the ratio of the number of different ways that the specified event can happen to the total number of possible outcomes.

Probabilities are expressed as a fraction between 0.0 and 1.0 or as a percentage from 0 to 100. A probability of 0 denotes an event that cannot happen, whereas a probability of 1.0 or 100 percent represents absolute certainty. This numerical description of the likelihood of an event's occurrence makes possible a numerical measure of many situations that could otherwise be expressed only instinctively or intuitively.

The most important principles of probability can be summarized in six general rules.

RULE 1.   Probabilities are, by definition, always between 0 and 1.0 (or 0 and 100 percent).

RULE 2.   The probability that an event will occur, plus the probability that it will not occur, equals 1.0 (or 100 percent).

RULE 3.   The probability that one or the other of two mutually exclusive events (events that cannot happen at the same time) will occur is equal to the sum of their individual probabilities of ocurrence.

RULE 4.   The probability that two or more independent events will happen simultaneously is equal to the product of their individual probabilities.

RULE 5.   The probability that a specified event will occur when its occurrence is contingent on another event taking place first is equal to the probability of the first event times the probability of the second events' occurring after the first has already happened.

RULE 6.   The probability of having either one or the other of two events happen when the events are not mutually exclusive is equal to the sum of their individual probabilities, less the probability of both happening at the same time.

Mathematical expectation combines the probability of an event's occurrence and the benefits to be gained by its occurrence. Expectation is calculated by multiplying the amount to be gained by a particular outcome by the probability of that outcome's occurrence. As such, mathematical expectation represents the average benefits that can be expected to result from a given situation over a period of time, taking into account both the gains from successes and the losses from failures.

PROBABILITIES LEAD TO THE MOST LIKELY PROFIT FROM NEW PRODUCT

By incorporating probability factors in the

$$\text{profit} = (\text{unit price} - \text{unit cost}) \times \text{volume}$$

relationship, the entire range of possible outcomes resulting from variations in the individual factors can be identified, and a project's profit expectations can be determined. Table 11.2 summarizes an example in which a product is to be manufactured. During its first year on the market, its average price will be determined by competitive factors and will range from $40 to $60 per unit with the probabilities shown in the table. Similarly, unit costs of

Table 11.2   Probabilities of Price, Cost, and Sales of a Product are Independent

Unit Price		Unit Cost		Sales Volume	
$	Probability	$	Probability	Units	Probability
40	0.2	30	0.3	1000	0.3
50	0.5	40	0.6	2000	0.4
60	0.3	50	0.1	3000	0.3

either $30, $40, or $50 may be incurred at the probabilities shown, and sales volume may vary according to outside, uncontrollable influences. In this example, the three independent variables—price, cost, and volume—are independent of each other.

The probability of occurrence of any given combination of the three independent variables (from the fourth rule of probability) is equal to the product of their individual probabilities. Since there are three different values for each of the three variables, a total of $3 \times 3 \times 3 = 27$ different outcomes are possible. These 27 outcomes are summarized in Table 11.3.

The outcomes range from the least profitable combination of the three variables (low unit price, high unit cost, high sales) to the most profitable (high price, low cost, high sales), with the results ranging from a $30,000 loss to a $90,000 profit.

Each outcome has its own probability of occurrence, as described before. By multiplying the profit to be made in a given situation by the probability of that situation occurring, the expected profit is found for each possible outcome. The expected profits for all 27 possible outcomes are added together to yield the project's expected profit: $26,000.

However, although the most likely profit is $26,000, there is no combination of price, cost, and volume that calculates to a profit of $26,000.

The value of such an approach is apparent, in that in addition to determining the most likely profit for the product, the entire range of possibilities is covered. There is, for example, a 2.0-percent chance that the project will lose money, and a 17.0-percent chance that it will just break even. The probability of making some profit, therefore, is 81.0 percent.

In Table 11.4, the data from Table 11.3 are summarized and rearranged in order of increasing profits. The second column shows the chances of earning (or losing) a specified amount. The third column gives the cumulative probability distribution, which is plotted in Figure 11.3 to illustrate graphically the range of possibilities associated with the proposed project.

In situations involving a large number of variables, or in which the range of possible values for the different variables is wide, the total number of combinations can quickly get out of hand. Just five variables, each having five possible values, will result in a total of 5^5, or 3125, different combinations. It is obviously impractical to attempt manual calculation of all the possible outcomes in such a case; the problem can easily be handled by a computer, especially one with an electronic spread sheet.

A computer can quickly calculate and summarize the complete range of possible outcomes, as in the preceding example. Or, it can produce essen-

Table 11.3 Expected Profits Associated with All Possible Combinations of Price, Cost, and Volume[a]

Price ($)	Cost ($)	Volume (units)	Profit ($)	Probability of Occurrence (%)	Expected Profit ($)
40	30	1,000	10,000	0.018	180
40	30	2,000	20,000	0.018	480
40	30	3,000	30,000	0.010	540
40	40	1,000	0	0.036	0
40	40	2,000	0	0.048	0
40	40	3,000	0	0.036	0
40	50	1,000	(10,000)	0.006	(60)
40	50	2,000	(20,000)	0.008	(160)
40	50	3,000	(30,000)	0.008	(180)
50	30	1,000	20,000	0.045	900
50	30	2,000	40,000	0.060	2,400
50	30	3,000	60,000	0.045	2,700
50	40	1,000	10,000	0.090	900
50	40	2,000	20,000	0.120	2,400
50	40	3,000	30,000	0.090	2,700
50	50	1,000	0	0.015	0
50	50	2,000	0	0.020	0
50	50	3,000	0	0.015	0
60	30	1,000	30,000	0.027	810
60	30	2,000	60,000	0.036	2,160
60	30	3,000	90,000	0.027	2,430
60	40	1,000	20,000	0.054	1,080
60	40	2,000	40,000	0.072	2,880
60	40	3,000	60,000	0.054	3,240
60	50	1,000	10,000	0.090	90
60	50	2,000	20,000	0.012	240
60	50	3,000	30,000	0.090	270
				1.000	26,000

[a] Sample calculation: P($40) = 0.20; P($30 cost) = 0.30; P(1000) = 0.30; Product = 0.018

Table 11.4   The Probability of Achieving Various Levels of Profit

Profit or Loss ($)	Probability of Occurrence (%)	Cumulative Probability (%)
(30,000)	0.6	100.0
(20,000)	0.8	99.4
(10,000)	0.6	98.6
0	17.0	98.0
10,000	11.7	81.0[a]
20,000	25.5	69.3
30,000	14.4	43.8
40,000	13.2	29.4
60,000	13.5	16.2
90,000	2.7	2.7

[a] Sample calculation: 2.7 + 13.5 + 13.2 + 14.4 + 25.5 + 11.7 = 81.0

tially the same results by drawing at random from a large number of possible combinations. The results of the random drawings will, as their number increases, cover the most likely occurrences with approximately the right frequencies, giving results comparable to detailed calculations in far less time.

STATISTICAL MEASUREMENTS

Precision in the anticipated levels of critical variables is unlikely. Often, only most likely value and some indication of minimum and maximum values can be anticipated.

For example, the time required to perform a certain task may vary between 90 and 110 percent of its average, and another task may vary between 80 and 120 percent. Some way of objectively expressing these variations from the average is desirable, so that the variations can be recognized and taken into account when project risks are being evaluated.

The average or mean value of a group of numerical data is found by summing the individual values and then dividing by the number of individual values in the group. In construction scheduling and estimating, the

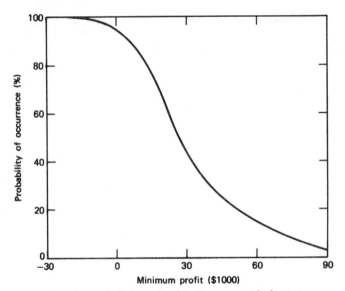

Figure 11.3. *The probability of achieving a specified minimum profit.*

mean or expected value for a specific task is often estimated from just the most optimistic, most pessimistic, and most likely values, as

$$v_e = \frac{a + 4m + b}{6}$$

where v_e is the expected value, a is the minimum value likely to be experienced under normal conditions, b is the maximum value that might be encountered, and m is the most likely value. If the data are normally or symmetrically distributed about their mean, v_e will be the same as the group mean. In a skewed distribution, the expected value will lean in the direction in which there is the greatest uncertainty or variation from the mean.

This formula can be used safely with most time or cost data. For example, if the most likely total cost of a certain manufacturing process were $200 per thousand units, but might vary between $175 and $250 under commonly encountered conditions, the cost to be used for estimating purposes would be:

$$v_e = \frac{175 + 4 \times 200 + 250}{6} = \$204$$

Such an estimate considers the possibility of both better and worse perform-ance than would normally be expected.

VARIANCE.   The statistical measure known as variance indicates the reli-ability of an estimate. Consistent with its name, variance is a measure of how big a spread exists between individual measurements and their average value. More specifically, variance is the average of the squares of the devia-tions of a number of observations from their mean value. If the variance is high, the degree of uncertainty associated with the estimate is high; a low variance indicates a high level of accuracy. For rough estimating purposes, variance is sometimes considered a function of just the extreme range of values—such as the most optimistic and the most pessimistic values in the preceding example—and can be approximated from the formula:

$$V = \frac{(b - a)^2}{6}$$

where V is the variance and b and a are the most pessimistic and most opti-mistic values, respectively. The variance in the manufacturing cost example is

$$V = \frac{(250 - 175)^2}{6} = \$156$$

This approach does not give a true measure of statistical variance; it does, however, give a reasonable indication of the relative reliability of var-ious estimates.

STANDARD DEVIATION.   Statistically, the standard deviation is the square root of the variance. If the data are distributed normally, two-thirds of the observations should fall within ± 1 standard deviation of the mean value, and 95 percent will be within ± 2 standard deviations. As with the variance, the smaller the standard deviation, the more reliable and predictable, and less risky, the estimate.

RISK, UNCERTAINTY, AND DEGREES OF IGNORANCE

The terms risk and uncertainty perhaps imply different meanings, yet both imply a degree of ignorance. Risk generally refers to situations in which the probability of occurrence of each possible outcome is either known from past experience or can be estimated, not just guessed at. Uncertainty, however, covers situations which are of a relatively unique nature and for which the probabilities cannot be stated.

Risk, then, implies less ignorance than does uncertainty. Stated otherwise, uncertainty is a risk that cannot be approximated.

Risks may be of either a technical or economic nature. Some risks must be taken, or are too inviting to pass up; some risks should be taken; and some risks are too great not to be taken. In any event, primary concern should not be with eliminating risks, but with selecting the right risks to be taken. The only way to avoid risk is to do nothing, which is actually the greatest risk of all in business. Actually, risks in business can never be avoided. Even a course of action that minimizes risk is not always desirable. There are several ways, though, that risk and uncertainty can be anticipated and handled so that their possible harmful effects can be minimized. For example, risks may be grouped to take advantage of the "law of large numbers," giving the effect of self-insurance. Or, responsibility for assuming risks may be delegated to an insurance company, which in turn consolidates its own risks by grouping the risks of many other individuals or firms.

Some types of risks can actually be controlled or protected against by instigating appropriate protective action. Accident risks, for example, can be controlled by safety measures, and the risks of fire and theft can be reduced by installing security controls.

In some cases, possible future difficulties can be anticipated on the basis of current information. Increasing the powers of prediction can help to avoid many unpleasant situations, either by changing events now, so that the unpleasant events never happen, or by directing the firm's operations along lines that will avoid the less desirable occurrences.

PROFITABILITY VERSUS RISK

The calculation of present worth or internal rate of return for a new venture should include three major elements: (1) pure interest, (2) compensation for management, and (3) compensation for risk.

The pure interest portion of the return is relatively easy to establish. It represents the return that could be realized by placing the available funds in some alternative, secure, interest-paying investment. This alternative investment might be certificates of deposit, treasury bills, high-grade bonds, or other investment media. In general, the rate of pure interest applicable to invested funds fluctuates between 8 and 10 percent, depending on the condition of the money markets.

Another 1 or 2 percent should be included in the project's target ROI as a reward for management's seeking out, evaluating, and reaching a decision on where the funds could best be placed.

Finally, the risk portion of the project return must be added. This is strictly a judgment factor and can range from 1 to 40 percent or more, depending on the particular project. Typically, a project having average risk should earn from 6 to 10 percent just on the basis of its risk, plus another 8–10 percent for pure interest and a 1–2 percent allowance for investment management. The average project, then, should earn from 15 to 20 percent after taxes, averaging about 18 percent, and paying out in about 7 years.

Table 11.5 summarizes these factors for five categories of projects, ranging from very low to high risk. Typically, these projects have profitability objectives of 11–41 percent after taxes, resulting in payout periods of between 12 and 2.5 years. Even under the best of conditions, it is difficult to justify projects earning less than 11 percent, or to find conventional projects returning more than 40 percent.

Venture capital firms investing in early-stage projects, however, generally hope to recover ten times their investment within a 5-year period—an annual return of 58.5 percent for these riskiest-of-all new ventures.

SUMMARY

Statistical techniques are valuable to the engineer in analyzing numerical data for use in predicting or anticipating the outcome of future events. Statistical analysis usually involves the development of frequency distribution data as a basis for probability calculations. Probability theory can be applied in predicting the likelihood of occurrence of uncertain events in numerical terms. The mathematical expectation associated with a variety of alternative situations can be developed by weighing both the probability of an event's occurrence and the potential benefits to be realized should it occur, thereby allowing all possibilities to be recognized and considered and minimizing the need for guesswork on the part of management. Risk and uncertainty are essential parts of business. Risks cannot be eliminated, but their harmful effects can be minimized by thoughtful planning, and business activities can be directed toward areas where risks are most strongly outweighed by potential profits. The profitability objectives for

Table 11.5 Typical Profitability Objectives for Ventures Having Different Levels of Risk

Risk Level	Interest (%)	Allowance for Management (%)	Risk (%)	Total Return (%)	Typical Return (%)	Typical Payout Time (years)
(1) Very low	8–10	0–1	1–3	9–14	11	12
(2) Below average	8–10	1–2	2–6	11–18	15	9
(3) Average	8–10	1–2	6–10	15–22	18	7
(4) Above average	8–10	1–2	12–20	21–32	25	4
(5) High	8–10	1–2	20–40	29–52	41	2.5

new ventures should be closely related to the risks involved, with high-risk projects requiring a correspondingly high rate of return and rapid payback. Typically, new projects should yield returns of 11–41 percent on invested capital, depending on the degree of risk associated with the venture and on the prevailing interest rate level.

12

Increased Profits From Financial Analysis

Before you organize you ought to analyze and see what the elements of business are.

Gerard Swope

Making a profit is what business is all about. Profits are adequate when they return to business owners the cost of their personally contributed resources, a reward for their entrepreneurship, and compensation for the risks involved.

The *adequacy* of profits is generally measured in terms of *profitability*, as a return on invested capital. However, profit analysis, as used here, deals primarily with the *magnitude* of profits and their relationship to other financial data.

201

FINANCIAL RATIOS FOR BUSINESS ANALYSIS

One of the best ways for an outsider to judge the effectiveness of a company's management is to analyze that company's financial statements. The firm's management can benefit greatly from the analysis by spotting weakness and potential trouble spots before they damage the firm's financial conditions beyond repair.

The two major financial reporting documents are the balance sheet and the operating (profit-and-loss or income) statement. The balance sheet shows what the company's financial resources are, as of a specific date, and the profit-and-loss statement describes—in dollars—the company's operations over a period of time. The absolute size of the figures on these documents is not particularly meaningful to the financial analyst, but rather the relationships between certain pairs of figures.

A corporation's balance sheet and profit-and-loss statements are "showpieces" prepared for widespread public distribution. The story told by these documents may appear to differ from what actually happened during the reporting period. Consequently, footnotes and special situations cited in these reports should be carefully examined and their effects noted on any figure used in ratio analysis. Even with these shortcomings, analysis of the reported figures can be very useful and productive.

The importance of the manner in which income taxes are accounted for cannot be overemphasized in financial analysis. A company's pretax profit is an unrealistic figure in business. The impact of taxes—both on the project in question and on the company's operation as a whole—must be carefully evaluated before an important investment decision is made.

Industry profits are often cyclical. Not all industries are profitable at all times. Even during periods of strong economic activity, the frequent occurrence of bankruptcies attests to the danger of assuming that profitability is uniformly distributed. Valid economic and financial data can be obtained on an industry-wide basis, and such data can be very useful as a basis for comparing the characteristics of individual companies. But only the data applying to a specific operation over a specific time can be used effectively in financial decision making.

Balance sheet and operating statement data for five selected corporations are summarized and presented in Tables 12.1 and 12.2. These data were taken from recent annual reports. Each of the five companies is among the leaders in its field, and would be considered a "blue-chip" investment.

In considering financial data, there is no such thing as an average com-

Table 12.1 Consolidated Balance Sheet Data for Selected Companies Adjusted[a] to One Million Dollars of Annual Sales

	Electrical Equipment Manufacturer	Retail Department Stores and Mail Order	Chemical and Drug Producer	Integrated Petrolem Company	Retail Grocery Chain
Assets					
Current assets					
Cash	34,000	18,000	38,000	31,000	12,000
Receivables	172,000	26,000	160,000	204,000	4,000
Inventories	175,000	184,000	174,000	90,000	76,000
Other	10,000	10,000	55,000	50,000	7,000
Total current assets	391,000	238,000	427,000	375,000	99,000
Property, plant, and equipment	198,000	87,000	445,000	795,000	92,000
Other assets	88,000	32,000	66,000	144,000	3,000
Total assets	677,000	357,000	938,000	1,314,000	194,000
Liabilities					
Current liabilities					
Accounts and notes payable	90,000	104,000	110,000	150,000	39,000
Other	159,000	33,000	42,000	49,000	18,000
Total current liabilities	249,000	137,000	152,000	199,000	57,000
Long-term debt	88,000	38,000	101,000	137,000	10,000
Other liabilities	47,000	3,000	27,000	19,000	7,000
Total liabilities	384,000	178,000	280,000	355,000	74,000
Stockholders' equity					
Common and preferred stock	54,000	14,000	206,000	115,000	13,000
Capital surplus and retained earnings	239,000	165,000	452,000	844,000	107,000
Total stockholders' equity	293,000	179,000	658,000	959,000	120,000
Total liabilities and stockholders' equity	677,000	357,000	938,000	1,314,000	194,000

[a] All figures are given in dollars.

Table 12.2 Consolidated Income Statement Data for Selected Companies[a]

	Electrical Equipment Manufacturer	Retail Department Stores and Mail Order	Chemical and Drug Producer	Integrated Petroleum Company	Retail Grocery Chain
Net sales	1,000,000	1,000,000	1,000,000	1,000,000	1,000,000
Cost and expenses					
Cost of goods sold	876,000	709,000	512,000	585,000	785,000
Other costs and expenses	44,000	222,000	338,000	285,000	185,000
Total costs and expenses	920,000	931,000	850,000	870,000	970,000
Net profit before income taxes	80,000	69,000	150,000	130,000	30,000
Income taxes	37,000	35,000	68,000	27,000	15,000
Net profit after income taxes	43,000	34,000	82,000	103,000	15,000

pany, just as no prospective customer or client is an average customer. The wide variations in the balance sheet and operating statement accounts for these five leading firms should offer ample confirmation of the nonexistence of an average firm.

Even when looking at a single company, it is not the absolute size of the figures on its financial statements that is particularly meaningful to the financial analyst. The relationships among the different figures, though, can sometimes be quite revealing.

A company's current financial standing (balance sheet) and past financial performance (income statement) can be interpreted most conveniently when the relationships among various balance sheet and profit-and-loss statement items are expressed as ratios. The calculated ratios for a company can be used to compare the company's performance with the averaged performances of other businesses of the same type. Similarly, a firm's progress or lack thereof can be shown by comparing several of its financial statements over an extended time.

TYPES OF FINANCIAL RATIOS.   Some ratios are derived wholly from the balance sheet. Others relate expense accounts to income accounts on the income statement. A third group combines items on the balance sheet with items on the profit-and-loss statement. These three types of financial ratios can provide management as well as outside analysts with the basic information needed for effective monitoring of the company's costs and finances.

Many different ratios can be derived from the balance sheet and profit-and-loss statement. In all, more than 30 different ratios are regularly compiled by various trade associations. Dun & Bradstreet, for example, publishes 14 key business ratios for more than 100 different business in retailing, wholesaling, manufacturing, and construction. Prentice-Hall and Robert Morris offer similar data.

RATIOS—SYMPTOMS OF PROBLEMS

Some ratios are important only in certain businesses; others are of more general application. The following ratios are of significance in almost every type of business enterprise.

Table 12.3 shows the financial ratios for the five selected companies. These ratios were calculated from the balance sheet data in Table 12.1 and the operating data in Table 12.2. The financial ratios shown in Table 12.3 differ substantially from those published by Dun & Bradstreet for the re-

Ratio	Balance Sheet	Income Statement
Net profit to net sales	—	x
Net profit to net worth	x	x
Net profit to working capital	x	x
Net profit to total assets	x	x
Net sales to net worth	x	x
Sales to working capital	x	x
Sales to fixed assets	x	x
Current assets to current liabilities	x	—
Cash to current liabilities	x	—
Fixed assets to net worth	x	—
Current liabilities to net worth	x	—
Total liabilities to net worth	x	—

spective industries. Even the industry averages vary widely from year to year. Ratios, similar to the financial data from which they are derived, apply only to a specific company at a specified time.

NET PROFIT TO NET SALES.   The net profit on sales, or profit margin, indicates to some extent a company's competitive strength or its vulnerability to a decrease in either its sales volume or its profits. Beyond a certain value, an increase in sales widens the profit margin, since fixed costs need not rise in direct proportion to sales. For the same reason, profits (expressed as a percent) tend to increase and decrease more rapidly than do sales. The profit margin also reflects the competitive situation within an industry. Grocery stores, for example, are noted for working on low profit margins. Among the individual companies surveyed, the grocery chain—as expected—showed the lowest margin, while the petroleum company was highest—indicating a year in which crude oil brought a higher price.

NET PROFIT TO NET WORTH.   The net profit/net worth ratio is perhaps the most important of all financial ratios, since it reflects the efficiency with which invested capital has been employed. A 10-percent return on invested capital is usually considered a bare minimum, and many well-managed companies earn more than 20 percent on their equity capital. The top 10 companies listed in Fortune's "500" generally earn better than 25 percent

Table 12.3 Financial Ratios for Selected Companies

	Electrical Equipment Manufacturer	Retail Department Stores and Mail Order	Chemical and Drug Producer	Integrated Petroleum Company	Retail Grocery Chain
Net profit to net sales (%)	4.3	3.4	8.2	10.3	1.5
Net profit to net worth (%)	14.7	19.0	12.5	10.7	12.5
Net profit to working capital (%)	30.0	33.7	29.8	58.5	35.7
Net profit to total assets (%)	6.3	9.5	8.7	7.8	7.7
Net sales to net worth (ratio)	3.4	5.6	1.5	1.0	8.3
Net sales to working capital (ratio)	7.0	9.9	3.6	5.7	23.8
Net sales to fixed assets (ratio)	3.5	8.4	2.0	1.1	10.5
Current assets to current liabilities (ratio)	1.6	1.7	2.8	1.9	1.7
Cash to current liabilities (%)	13.6	13.1	25.0	15.6	21.0
Fixed assets to net worth (%)	97.6	66.5	77.8	97.9	79.1
Current liabilities to net worth (%)	85.0	76.6	23.1	20.7	47.5
Total liabilities to net worth (%)	131.0	99.4	42.6	37.0	61.6

on their invested capital. All five of the sample companies showed better than a 10-percent return on net worth, with the department store operation realizing 19 percent. The return on net worth figure for a company gives a rough indication of what constitutes a *minimum* acceptable return on a new investment for that company. Certainly, any investment proposal indicating a return lower than what the company is presently earning can not be considered a particularly attractive investment opportunity.

NET PROFIT TO WORKING CAPITAL.   Working capital represents the equity of owners in the company's current assets: the difference between total current assets and total current liabilities. This margin represents the "cushion" available to the business for carrying receivables and for financing day-to-day operations. The ratio of net profits to working capital is useful in measuring the profitability of firms whose operating funds are provided largely through borrowing, or whose permanent capital is unusually small in relation to its volume of sales. Four of the five selected companies had profit/working capital ratios within a narrow range, with only the petroleum company falling outside the 29–36 percent area. This ratio is more useful in comparing consulting or service firms.

NET PROFIT TO TOTAL ASSETS.   The ratio of net profits to total assets is closely related to the net profit/net worth ratio, except that here the theory is that return on investment should be measured in terms of *all* capital employed in the business—whether supplied in the form of equity or debt—and not in terms of equity interest only. All five companies surveyed earned between 6.3 and 9.5 percent on their total assets, even though their debt capital (as indicated in their balance sheets) may vary by as much as a factor of 10.

SALES TO NET WORTH.   This ratio measures the rate of capital turnover, showing how actively the firm's capital is being put to work. If capital is turned over too rapidly, accounts payable and short-term debts are apt to build up at an excessive rate; if capital is turned over too slowly, funds become stagnant and profitability suffers. The importance of capital turnover is vividly illustrated by comparing the profitability of the grocery chain and the chemical producer in the example. Although the chemical firm achieved a profit margin more than five times as high as the grocery chain, its rate of capital turnover was less than a fifth that of the grocery. Since the return on net worth is equal to the profit margin multiplied by the capital turnover rate, both attained the same 12.5 percent return on their net worth.

SALES TO WORKING CAPITAL.   The rate of working capital turnover can highlight a financial problem if it is either very low or very high. If the sales/working capital ratio is high, or growing, the business may owe too much, relying on credit as a substitute for an adequate margin of current operating funds.

SALES TO FIXED ASSETS.   This ratio is less significant in itself than when compared with the same ratio for previous years. Such a comparison shows whether or not the funds used to increase productive capacity are being spent wisely. If comparable sales increases have failed to accompany sizable investments in fixed assets, then poor asset utilization is indicated. This ratio is seldom significant for consulting or service firms, which can literally lease everything they use.

CURRENT ASSETS TO CURRENT LIABILITIES.   This ratio—also known as the "current ratio"—is one of the most commonly used ratios in balance sheet analysis. It indicates the margin of protection for short-term creditors, a 2:1 ratio being considered a standard. In general, the more liquid the current assets, the less marign needed to cover current liabilities.

CASH TO CURRENT LIABILITIES.   The ratio of cash and its equivalents (marketable securities) to current liabilities—sometimes referred to as the "liquidity ratio"—is a supplement to the current ratio, indicating the firm's ability to meet an immediate need for cash without borrowing or selling stock.

FIXED ASSETS TO NET WORTH.   A firm's tendency to overinvest in fixed assets can often be identified by this ratio. A high ratio of fixed assets to net worth results in heavy depreciation and interest burdens, which mean high fixed costs and serious profit problems.

CURRENT LIABILITIES TO NET WORTH.   The ratio of current liabilities to net worth provides a means of evaluating a company's financial condition by comparing what is owed with what is owned. A high ratio—any value over 0.80—indicates that the firm is overly dependent on its creditors.

TOTAL LIABILITIES TO NET WORTH.   Whenever the total liabilities/net worth ratio exceeds 1.00, it indicates that the creditors have a greater equity in the firm's assets than do the owners. Such top-heavy liabilities make the business extremely vulnerable to any unexpected contingencies and severely restrict the management's flexibility. Lenders are known to frown on clients who do not control this ratio.

In summary, the final measure of management effectiveness can be found in a firm's balance sheet or profit-and-loss statements, where financial and operating strengths and weaknesses can be objectively identified and analyzed.

GRAPHING THE PROFIT/VOLUME (P/V) RATIO—A SPECIAL CASE OF BREAK-EVEN ANALYSIS

The same techniques are employed in analyzing profits as in breakeven analysis, which is simply a case of profit analysis with zero profit. Graphical presentation of the data is again most useful and easily understood.

The relationship between profit and volume is sought in the typical profit analysis study, volume being the independent variable and profit the dependent variable. Very few data are required to construct a simple P/V chart, as shown in the following condensed operating statement for a company:

Total sales		$1,000,000
Cost of operations		
Variable costs	$670,000	
Fixed costs	$230,000	
Total costs		$ 900,000
Net profit before income tax		$ 100,000

These data can be handled in generally the same way as in a conventional breakeven analysis. Figure 12.1 shows a P/V chart developed from this information. The vertical axis is divided into two sections, the top part representing net profit and the lower segment depicting net loss. The horizontal axis is drawn through the zero profit point on the vertical axis, representing the breakeven point. Units along the horizontal axis measure the company's total sales volume.

This P/V chart indicates the net profit or loss associated with any given level of sales.

At zero sales, the net loss is equal to the amount of the overhead or fixed

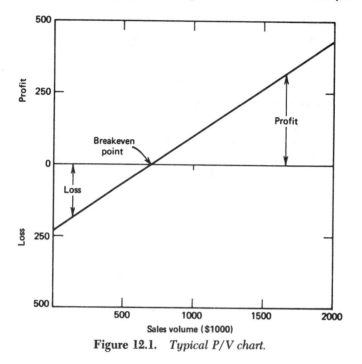

Figure 12.1. *Typical P/V chart.*

expenses. The P/V line intersects the vertical axis at a net loss of $230,000, which is incurred at zero sales volume. According to the operating statement, the company's net profit at a sales volume of $1,000,000 was $100,-000; therefore, those figures fix one point on the graph. A line connecting these two points gives the amount of profit or loss corresponding to any chosen level of sales. That the line is straight assumes that all other factors—such as the markups on direct costs, the amount of variable cost per dollar of sales, and the amount of overhead expense—remain the same. The point at which the P/V line intersects the horizontal axis—at $697,000 in this case—is the breakeven point. The area enclosed by the P/V line and the two axes to the left of the breakeven point represents net losses; and the area bounded by the P/V line and the axes to the right of the breakeven point describes the firm's profits.

Other things being equal, profits can be interpolated or extrapolated easily for any desired or expected sales volume. For this operation, the $1,000,-000 sales brought in $100,000 in profits. A sales volume of $1,500,000 should therefore yield $265,000 in profits, and a further increase in sales to $2,000,-000 will result in profits of $430,000. A reduction in sales to the $500,000 level will result in a net loss of $65,000. The slope of this line is such that each $100,000 change in volume causes a corresponding $33,000 change in profit.

APPLICATIONS OF THE P/V RATIO

The P/V ratio affords a convenient way to measure the firm's profit margin—the difference between total sales and total direct costs. In the preceding example, each $100,000 increment of sales was accompanied by a marginal profit of $330,000. The P/V ratio is, therefore, 0.33 and can be used to compute the firm's breakeven point and net profits.

FINDING THE BREAKEVEN POINT.   In the example, divide the total fixed cost by the P/V ratio:

$$\text{Breakeven Point} = \$230{,}000/0.33 = \$697{,}000$$

CALCULATING NET PROFIT.   The profit at any given sales volume can be found by multiplying the sales by the P/V ratio and deducting fixed expenses:

$$\text{Net Profit at } \$1{,}200{,}000 \text{ Sales} = (\$1{,}200{,}000 \times 0.33) - 230{,}000$$
$$= 396{,}000 - 230{,}000 = \$166{,}000$$

OTHER FACTORS ALSO AFFECT PROFITS

As long as the relationships between overhead expenses, direct costs, and sales volumes remain constant, the foregoing applications of the P/Vratio are valid. However, the entire profit situation changes whenever the markup, and consequently the variable cost rate, is changed. The following formula is useful in identifying the impact of changes in fixed costs, markups, and sales volumes on a firm's profits.

$$P = S \left(\frac{M}{100 + M} \right) - F$$

where P is the net profit at sales volume S, M is the average percentage markup on direct costs, and F is the total amount of overhead or fixed expenses.

Referring to the same example, the average markup can be found by dividing the total sales dollars by the total direct (variable) costs: $1,000,-000/$670,000 = 1.493$, indicating an average markup of 49.3 percent. The net profit resulting from using this markup at the $1,000,000 sales level is computed as follows:

$$P = 1,000,000 \times \left(\frac{49.3}{100 + 49.3} \right) - 230,000$$

$$= (1,000,000 \times 0.33) - 230,000 = \$100,000$$

Changing the percentage markup on direct costs to 60 percent while keeping the other factors constant gives a profit of

$$P = 1,000,000 \times \left(\frac{60.0}{100 + 60.0} \right) - 230,000$$

$$= (1,000,000 \times 0.375) - 230,000 = \$145,000$$

In this example, increasing the percentage markup from 49.3 to 60.0 percent increases the profit from $100,000 to $145,000. Thus, a 20-percent increase in the markup results in a 45-percent increase in profits at these levels. A change in the amount of overhead expense, of course, brings about a corresponding change in profits at any prescribed level of sales and markups; each dollar decrease in overhead raises profits by a dollar.

THE LAWS OF PROFITS

The relationships between profits, sales, overhead expenses, direct costs, and percentage markups lead to some general observations and conclusions. Briefly, these "laws of profits" are as follows.

1.   *All* costs—not just out-of-pocket direct costs—must be included when analyzing a firm's profit picture.

2.   Any change in fixed costs changes the net profit by a like amount, in the opposite direction.

3.   A given percentage change in fixed costs changes the firm's break-even point by the same percentage and in the same direction, providing other cost relationships hold steady.

4.   A change in the rate at which variable costs are incurred changes both the breakeven point and the net profit earned at a specified sales vol-

ume, if fixed costs remain constant. Increasing the variable expense rate raises the breakeven point and lowers profits; decreasing the rate lowers the breakeven point and increases profits.

5.   An increase in the percentage markup applied to direct costs lowers the breakeven point and increases the profits associated with any sales volume. Decreasing the markup has the opposite effect.

6.   A change in both fixed costs and the variable cost rate has a significant effect on profit if they both change in the same direction; there is less impact if they move in opposite directions.

THE STAKES GO UP WITH OUTSIDE FINANCING

In engineering projects, the method of financing is often ignored in the feasibility study or preinvestment analysis. When the DCF approach is taken, it is simply assumed that, if the necessary capital can be obtained at a rate less than the indicated IRR, the project can be undertaken at a profit. However, if the project's expected IRR is less than the company's cost of money, the project should be avoided.

No amount of outside financing can turn a basically unprofitable venture into a profitable one. But when the capital necessary to undertake a new venture can be obtained from outside sources at a rate less than what the project will return during its economic life, the IRR can be substantially amplified. Perhaps the most important single financial tool available to investors and businessmen is the leverage afforded by using other people's money (OPM). If, for example, a $100,000 investment were available that would yield a 15-percent annual return, and if the investor were able to borrow $75,000 at a 9-percent interest rate, his return on the $25,000 invested from his own pocket would be $(0.15 \times 100,000) - (0.09 \times 75,000) = \$8250$/year. This amounts to a return of $8250/25,000 = 33$ percent. In this case, borrowing 75 percent of the required capital provides leverage that more than doubles the percentage IRR. If this same investor had $100,000 of his own money, he could seek out three more investments under comparable terms and end up with an annual return of $33,000—compared with only the $15,000 he would have realized had he attempted to finance the entire investment without outside sources of funds.

Leverage works in the other direction too. Should the project fail to earn at the expected rate, and losses are incurred in a highly leveraged position, the effect can be devastating. Even earning just a little less than the cost of the borrowed money can have a damaging impact on a company's financial strength.

If the investor who borrowed 75 percent of his capital at a 9-percent interest rate managed to earn only 6 percent on the total amount invested in the project, he would end up with an out-of-pocket loss of $750; his project would earn $6000 while paying out $6750 in interest charges. If the investor had a more highly leveraged position—for example, say in the 80–90-percent range—the impact would be far greater.

Figure 12.2 shows the effect of different levels of borrowing in this case, with all other conditions remaining the same. As indicated in the figure, the return on owner's equity goes from 15 percent when no outside borrowing is used, to 21 percent when half the total capital is borrowed, and up to 69 percent when 90 percent of the money is obtained by borrowing. Conversely, should the project earn only 6 percent on total capi-

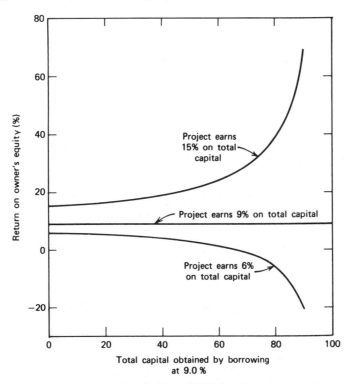

Figure 12.2. *Example of effect of OPM on investment return.*

tal, the percent return on the owner's equity would be accelerated in a negative direction.

MATHEMATICAL RELATIONSHIPS. The change in an investment's return brought about the leverage of borrowed money can be approximated by using the following formula:

$$R_p = \frac{R_0 - pI}{1 - p}$$

where R_p is the return when the portion p (expressed as a decimal) of the required capital is borrowed at a specified interest rate I. The return that would be earned if all money were supplied internally, without borrowing, is expressed as R_0.

Thus, if a project were expected to earn a net return of 15 percent annually without benefit of outside financing, but if 80-percent financing could be arranged at a 9-percent interest rate, the new return for the project would be:

$$R_{0.80} = \frac{15 - 0.80 \times 9.0}{1 - 0.80} = \frac{7.8}{0.2}$$

$$= 39 \text{ percent}$$

This formula is approximate only because it does not consider the timing of principal and interest repayments on the borrowed money, nor does it take into account the effect of interest payments on income taxes and, consequently, on cash flow patterns. It does, however, give a good, quick indication of the financial benefits of using OPM.

Even when the cost of borrowed money is greater than the indicated return for the project, an investment still might be justified if there were no other way to go ahead with it, and if the return were still adequate. A project starting with a return of 10 percent and financed one-half through borrowed 12-percent money still yields:

$$\frac{10 - (0.5 \times 12)}{1.0 - 0.5} = \frac{4.0}{0.5} = 8 \text{ percent}$$

In this situation, even though the project's return is reduced by the necessity to use OPM, the return might still be high enough to warrant consideration. If the project could only be undertaken by borrowing money, then borrowing under even these conditions might be acceptable.

REPAYMENT OF BORROWED MONEY

The timing of the borrowing of money and especially the timing of principal repayments are also critical factors in leveraging an investment's IRR. This is particularly noticeable in real estate transactions, in which property may still carry a substantial mortgage when it is sold.

For example, consider a tract of land purchased for $40,000, of which $10,000 is paid down and the $30,000 balance is spread in equal payments over a 5-year period at 16-percent interest. The annual repayment schedule is:

Year	Principal	Interest	Total
0	10,000	0	10,000
1	6,000	4,800	10,800
2	6,000	3,840	9,840
3	6,000	2,880	8,880
4	6,000	1,920	7,920
5	6,000	960	6,960
Total	40,000	14,400	54,400

If the property were sold for a cash price of $45,000 at the end of the first year, the total payments to date would consist of $16,000 in principal and $4800 in interest. Part of the $45,000 received for the land would be used to pay off the $24,000 balance, leaving $21,000. The net capital gain on the property would be $5,000, or 31 percent of the $16,000 actually invested. This gain is brought about by an increase in land value of just 12.5 percent, coupled with the leverage of OPM.

If the property had been sold just before the first annual payment became due, the $45,000 would have paid off the $30,000 balance and the $4800 interest, leaving $10,200 and a clear $200 gain on the 1-year investment of $10,000. Again, $5000 would be the capital gain and $200 would be tax-deductible as interest paid.

From a profitability standpoint, as long as an investment earns more than the capital costs, the investor benefits from the leverage of OPM. Therefore, to attain the highest possible return, the principal repayments should be deferred as long as possible. In summary, borrow as much as possible, as cheaply as possible, for as long as possible; pay it back as late as possible, and make a profit on it.

THE MORE BORROWED, THE MORE THE POSSIBLE PROFIT

There are basically four ways an investment can be financed and repaid: (1) cash, (2) equal principal payments, (3) equal annual payments, and (4) interest only, with the principal repaid in the final year.

CASH ON THE BARRELHEAD.   The simplest and least profitable way to handle a new capital investment is to use your own money. At one time, many businesses were conducted on this basis, and a few still are. If large cash reserves are available, and there is not a better use for them, then this approach at least keeps them from sitting idle. Table 12.4 shows the cash flows associated with a cash investment; the proposed venture is expected to generate operating profits of $6000/year before taxes and $3000/year after taxes. The entire $10,000 investment is made initially, followed by 5 years of cash inflows. Since there are no interest payments, the net cash inflows are the same as the net profits after taxes. This investment, set up in this way, yields an IRR of 15.2 percent on a DCF basis.

EQUAL PRINCIPAL PAYMENTS.   Intermediate-term business loans are often repaid in equal principal installments plus accrued interest on the unpaid balance. This approach is easy to handle, interest calculations are simple, additional principal payments can be made without disturbing a preset repayment schedule, and the lender can see the loan balance being reduced at regular intervals. Large land purchases are often owner-financed in this way, with payments made either annually or semiannually. Table 12.5 shows an example of this approach, using the same proposed venture as in the previous example. However, only $2000 is applied as a down payment at the beginning, and the $8000 balance is spread out in $1600 payments annually over the 5-year period. Since interest charges (at 16%) are computed on the unpaid balance, interest costs decrease each year. The net cash flows (the net profit after taxes, less principal payments) increase each year. The IRR earned by financing the project this way increases to 21.4 percent.

EQUAL ANNUAL PAYMENTS.   Mortgages on commercial and residential buildings are generally set up on an amortization schedule, where a uniform periodic (usually monthly) payment covers both principal and interest. The principal portion increases with each payment, while the interest portion decreases as the unpaid balance declines. Borrowers usually find the equal payments of amortized loans easier to include in their financial planning, although their equity builds slower than with equal principal payments. In

Table 12.4 Cash on the Barrelhead

Year	Operating Profit	Interest	Net Profit Before Taxes	Net Profit After Taxes	Principal Payments	Net Cash Flow
90	0	0	0	0	10,000[a]	(10,000)
1	6,000	0	6,000	3,000	0	3,000
2	6,000	0	6,000	3,000	0	3,000
3	6,000	0	6,000	3,000	0	3,000
4	6,000	0	6,000	3,000	0	3,000
5	6,000	0	6,000	3,000	0	3,000
Total	30,000	0	30,000	15,000	10,000	5,000

[a] Not, strictly speaking, a principal payment, this is really the basic investment—in cash.

Table 12.5 Equal Principal Payments

Year	Operating Profit	Interest (16%)	Net Profit Before Taxes	Net Profit After Taxes	Principal Payments	Net Cash Flow
0	0	0	0	0	2,000	(2,000)
1	6,000	1,280	4,720	2,360	1,600	(760)
2	6,000	1,024	4,976	2,488	1,600	888
3	6,000	768	5,232	2,616	1,600	1,018
4	6,000	512	5,488	2,744	1,600	1,144
5	6,000	256	5,744	2,872	1,600	1,272
Total	30,000	3,840	26,160	13,080	10,000	3,080

Table 12.6 Equal Annual Payments ($2443)

Year	Operating Profit	Interest (16%)	Net Profit Before Taxes	Net Profit After Taxes	Principal Payments	Net Cash Flow
0	0	0	0	0	2,000	(2,000)
1	6,000	1,280	4,720	2,360	1,163	1,197
2	6,000	1,094	4,906	2,453	1,349	1,104
3	6,000	877	5,123	2,562	1,566	996
4	6,000	627	5,373	2,687	1,816	871
5	6,000	377	5,663	2,831	2,106	725
Total	30,000	4,215	25,785	12,893	10,000	2,893

Table 12.7 Interest Only—Principal in the Final Year

Year	Operating Profit	Interest (16%)	Net Profit Before Taxes	Net Profit After Taxes	Principal Payments	Net Cash Flow
0	0	0	0	0	2,000	(2,000)
1	6,000	1,280	4,720	2,360	0	2,360
2	6,000	1,280	4,720	2,360	0	2,360
3	6,000	1,280	4,720	2,360	0	2,360
4	6,000	1,280	4,720	2,360	0	2,360
5	6,000	1,280	4,720	2,360	8,000	(5,640)
Total	30,000	6,400	23,600	11,800	10,000	1,800

most amortized mortgages, the payment schedule must be strictly adhered to, or the balance paid off in a lump sum. Table 12.6 describes the same investment as described previously, repaid in equal annual amounts (principal plus interest = $2443). Here, net cash inflows are highest during the early years, thereby bringing the IRR up to 44.5 percent.

INTEREST ONLY.   The situation in which principal repayment is delayed until the project's final year, and is then paid off in a lump sum, is best from the borrower's profitability standpoint. Until then, only interest payments are made. Corporate and municipal bonds are set up this way, paying interest annually or semiannually until they are retired at a specified future date. As in the case with an amortized loan, early retirement can usually be made only by paying off the entire debt balance in a single amount. Table 12.7 illustrates this type of repayment schedule, again with a $2000 down payment required at the beginning. The $8000 principal balance is carried until the final year, when it is paid off in one lump. Meanwhile, interest is paid annually (computed at 16 percent of the $8000 unpaid balance), and cash flows are uniformly high until the final principal payment is made. Here, the IRR is an impressive 102.5 percent/year.

These four examples differ only in the way principal and interest payments affect cash flows, yet the resulting IRRs vary by a factor of almost 7, from 15.2 to 102.5 percent. In each case, higher interest payments, reflecting deferred principal payments, bring higher investment returns.

SUMMARY

Financial analysis provides the means by which a firm's financial performance can be objectively evaluated. Several important financial ratios—developed from balance sheet and operating statement data—can be used to compare a company's performance with others in the same field, or with itself over a period of time. The profit-sales and profit-net worth relationships are particularly significant in showing a company's competitive strength and economic efficiency. The profit/volume ratio is also useful in anticipating the changes in profits associated with changes in sales; this ratio can be used to calculate breakeven points and profits under varying conditions. Many factors affect a firm's profits and profitability: changes in fixed costs; the rate at which variable costs accrue; prices, as determined by percentage markups on direct or variable costs; the sources and costs of outside funds employed; and the timing and methods of repaying borrowed money. Considerable leverage can be gained by using outside financing whenever the cost of borrowed money is less than the proposed venture is expected to earn. The project's overall return in such a case can be further enhanced by delaying repayment of the principal amount for as long as possible.

13

Measuring
Investment
Performance

Profitability is the sovereign criterion of the enterprise.

Peter Drucker

One of the important differences between a consulting engineer and an engineer in industry is apparent in the follow-up of a recommended investment. The engineer in industry may have to live with his decisions and recommendations for many years (no matter how embarrassing they may be); the consulting engineer may never even find out whether his recommendations were followed. If they were followed, he will probably never know just how closely the results of his recommended projects came to the estimates and forecasts upon which they were based. In such situations, the engineer can do little to improve his performance in project evaluation, planning, and forecasting. Without the guidance of a project follow-up, the results of existing techniques are not known, and the engineer lacks incentive to develop new or better techniques.

That missing guidance can come from a postinvestment appraisal: a comparison of the actual results of an investment decision with the results originally estimated in justifying the investment decision in the first place. Thus, the postinvestment appraisal offers the engineer an opportunity to objectively appraise the quality of his own work by showing to what extent investment proposals have achieved the results forecast for them.

Other equally descriptive terms sometimes used for the postinvestment appraisal include postinvestment audits, postcompletion reviews or audits, performance reviews, experience reports, or postmortems.

THE POSTINVESTMENT AUDIT

Benefits for all concerned

The characteristics of a postinvestment audit depend largely upon the purposes for which it is made. For many projects, the engineer's first "audit" occurs when bids are opened for the initial construction work. At this point, a hasty review of the entire investment proposal may be necessary. If that review includes some extra care in planning the construction phase of a new project, many unpleasant surprises can be avoided.

Comparing a capital cost estimate to contractors' bids and actual costs, though, is only one of many useful purposes to which postinvestment appraisals can be put. Other equally important reasons for making these audits include:

1. Improving the evaluation of future projects.
2. Insuring realistic estimates by sponsors of new projects.
3. Improving the operating performance of completed projects.
4. Providing management with more accurate estimates of the success of new projects.
5. Identifying areas where the investment analysis process needs improvements or remedial action.
6. Determining the soundness of original assumptions.
7. Disclosing inadequacies in the form or content of requests for capital appropriations.

If a postinvestment appraisal program is carried out on a very imper-
sonal basis, and if great care is exercised to avoid placing unwarranted
blame on individuals, the results will be valuable to everyone concerned. So
many different events and decisions may have preceded the investment decision that it is extremely difficult to pinpoint the specific factors that affected its ultimate profitability. If postinvestment appraisals are properly
administered and used in a systematic and thorough manner, however, they
can greatly enhance the engineer's knowledge and understanding of what
happened in the past, and thus improve his capabilities for performing
under similar conditions in the future.

Opportunities for better estimates or more business

According to surveys of major United States industrial firms, nearly 90 percent of the companies employ performance reviews on selected new projects, beyond just an accounting review of actual versus estimated capital
expenditures.

For private or industrial firms, any project that merited the name, and
for which even a rudimentary economic analysis was run, merits the time
and cost of a postmortem. The best projects to postaudit are those entered
into directly for profit, in which both operational and financial performance
can be measured. Typical projects include new product lines, replacement
of equipment, and expansion of facilities. For municipal or public organizations, postinvestment appraisals are particularly important on projects financed through revenue bonds, service charges, or other revenue sources
specifically related to the project.

The key to obtaining maximum benefits from the postinvestment appraisal is to evaluate the important factors involved in initially justifying
the project, especially those to which the project's economic success was
most sensitive. A project's IRR, for example, is usually most sensitive to its
initial capital outlay and its gross operating profits in its early years. These
are the logical factors to scrutinize in any postinvestment reviews of projects entered into for profit.

Since the consulting engineer's responsibility for a new project generally
ends shortly after the startup period is over, his interest in the project may
unfortunately end there too. As noted, this is one of the main differences
between the consulting engineer and his in-house counterpart. There is one
difference that can easily justify the effort of an appraisal by the consulting

engineer—use of the results of the appraisal in developing business with new clients.

THE AUDIT FACES PROBLEMS AND LIMITATIONS

The industrial firm monitoring the results of its own decisions on its own terms is faced with several problems in administering a post-investment appraisal program. The consulting engineer, in addition to having all the conventional problems, is also apt to be faced with some problems in gaining access to the data necessary for a realistic audit.

One of the main problems from the engineer's standpoint is that although the audit measures the results of what was done, it does not indicate whether or not the original design was the best possible. For example, an audit of a sewage treatment facility might show that the trickling filter design criteria were correct, the construction costs were accurate, the operating costs came out as expected, the organic and hydraulic loadings were properly anticipated, and the effluent quality was predicted. But the audit does not reveal whether the trickling filter plant that was installed was preferable to alternatives available when the investment decision was made. Moreover, obtaining relevant operating data, especially from municipalities, may turn out to be strictly a do-it-yourself project for the engineer.

Internal problems abound within companies or departments whose investment programs are being audited. Corporate "in-fighting," a general lack of incentive, and extreme reluctance to revive old problems that would sooner be forgotten pose serious barriers to the development of good post-investment audits. For all those reasons, it is difficult to come up with good investment reviews, regardless of how valuable they might be.

Some of the other problems associated with postinvestment appraisals include a tendency toward complacency when performance is good and a corresponding reluctance to expose the facts when performance is poor. Also, obtaining the appropriate financial data may be a problem in that the data in a company's accounting system may not be directly identifiable with specific projects, and some costs, such as overheads and administrative expenses, may not be easily or rationally allocable. Even in organizations that have a project cost accounting system, project managers have been known to allocate direct charges according to which project is within budget, rather than to the project for which the money was spent.

The most serious disadvantage of the postinvestment appraisal is the tendency of auditors to look for things that went wrong. Thus, the audit "grades" the performance of the project economist on a scale that runs from some negative value up to zero. Projects are usually budgeted so carefully that a major technical breakthrough would be required to complete a project before the planned date or under budget. The result is an apprehensive economist who sees the auditor as a means to career injury rather than someone whose report could mean a promotion to project manager.

MAKE THE APPRAISAL AS PART OF THE PROJECT PLAN

As with any other aspect of a project, the postmortem will be more effective and useful if it is included in the overall plan of the project. Such an inclusion leads to four major considerations in the administration of an effective postinvestment appraisal program. They are:

1. Assignment of responsibility and cost.
2. Selection of projects.
3. Timing of audits.
4. Reporting audit results.

RESPONSIBILITY.   Responsibility for conducting an audit of a completed project depends primarily upon the reasons for performing it. If the main purpose of the audit is to help insure realistic estimates by checking up on the accuracy of the firm's estimators, then responsibility should be assigned to a group operating independently of those who prepared the original estimates. But if the purpose of the audit is to improve future evaluation techniques, then the same people who prepared the estimates should also conduct the audit.

Deciding whose budget will cover the cost of the audit can be ticklish, especially if no one other than the controller is interested in the results. That decision is easiest to implement if it is made before the project starts.

SELECTION OF PROJECTS.   The selection of projects for review is usually determined by some set dollar amount, extending from a minimum between $5000 and $100,000. The projects selected should be those expected to yield the greatest amount of information for use on similar future projects.

TIMING.   Timing of the postinvestment appraisal is extremely critical, if the best possible results are to be obtained. If the performance review is conducted too soon following the expenditure, there will be no accurate reflection of the project's long-range success or failure. But if the audit comes too late, the data are harder to obtain, and the potential benefits of the review may be lessened because of changes in personnel and procedures. From the consultant's viewpoint, calling on a client to review an unsuccessful project's results may prove an embarrassing experience. Such calls, however, can help consultants learn what they did wrong and how to avoid repeating the errors. In addition, such calls can help disclose and then repair damage to the client-consultant relationship that may have occurred during the more critical phases of the project. On the other hand, when that relationship has been close and productive, an audit can lead to improved approaches to new problems.

If a project is planned as a multiphase effort, the client or the project manager may prefer a similar plan for the audit. In fact, a phased audit can provide the project manager with an additional means of control.

AUDIT REPORTS.   The written record of an audit of a completed project requires a flexible format, but usually consists of two main sections. The first section includes a review of the original objectives of the project, general comments regarding the conditions encountered during the time period over which the review was made, and a brief statement concerning the outlook for the future. The first section of the performance report, then, is primarily narrative and subjective in nature. The second section contains comparisons of profit, sales, manufacturing or production costs, volume handled, or whatever other parameters were used in the initial study with estimates for the originally proposed project. Adjustments can be made in the original estimates to reflect changes in wage rates, materials prices, and general economic conditions.

INTERPRETATION OF RESULTS

After all the information has been collected that will make possible a meaningful comparison of the actual results of a project with the estimates upon which the project was originally based, the next step is to interpret any significant differences between actual and estimated results. In doing this, some distinction should be made between estimates based on formal forecasting methods and those based on individual judgment.

If the estimates were based on formal techniques that have been developed from past experience, studying a sample of the estimates they produced can test the validity of the techniques. It is almost equally easy to determine whether the forecasting method was properly applied in a specific case.

When estimates were based on an individual's judgment, though, there is little to be learned by comparing an actual result with an estimate in a specific situation. More can be learned in this case by comparing the factors that the individual(s) considered important in making the original estimate with the factors that actually influenced the project's performance—essentially, an after-the-fact sensitivity analysis. If an individual's judgment is to be appraised, he can be viewed as a "black box" by comparing the actual and estimated results on several projects in which his judgment was exercised.

Whenever possible, interpretation of the comparisons between actual and estimated results should be based on groups of similar projects, rather than upon single projects. If the purpose of the postinvestment audit is to improve forecasting methods or to insure more realistic estimates, considering more projects will provide better results.

In all comparisons, the estimates and the bases on which they were made should be explicitly defined and available at the time of the audit. All factors considered in the judgments made in connection with the estimates should be clearly stated, and all forecasts and assumptions used in the final estimate should be recorded. Only then is it possible to determine which of the preliminary forecasts or assumptions caused the final forecast to vary from the actual results.

Economic justification of capital expenditures encompasses two broad areas. First, it requires the development and application of reliable methods for determining the financial and/or operating success of new capital investments—methods for screening new projects and for evaluating proposed investments. Second, it is concerned with a continuing review of the investment decision process through comparisons of actual results with the assumptions and data upon which the expenditures were initially justified.

CASE HISTORIES USUALLY REFLECT OVEROPTIMISM

The following three case histories have been drawn from the experience of manufacturing firms in different areas and involve different types of projects. The first two examples deal with new manufacturing facilities built to produce new products; the third example describes the results of an acquisition.

These three examples are not spectacular; they do, however, demonstrate the usefulness of postinvestment analyses. In two cases, actual profitability was lower than that predicted in the preinvestment evaluation. Disappointments of that sort are the rule. The unpleasant surprises can arise from:

1. A generally overoptimistic project economist.
2. The estimated selling price is unrealistically high.
3. The enthusiasm of the market for the product is overestimated, and an unattainably high sales volume is projected.
4. The time required to penetrate new markets is underestimated.
5. Reactions of competitors are not currently anticipated.
6. The time required to solve production problems is underestimated.

Realistic assessments of markets, prices, and production costs—entailing in-depth market and cost research—are necessary components of a good investment analysis. The cost engineer must also guard against being overly optimistic, despite the enthusiasm shown by a new project's sponsors. Investment or capital requirements are the most reliable part of most evaluations, since the cost engineer or estimator has generally had enough experience in that area to develop reasonably accurate figures. Market research, however, is a field calling for special qualifications in areas in which most engineers are less experienced.

EXAMPLE 1:   Facilities to Manufacture a New Chemical Product

This project dealt with a new development that was based on a discovery by the company's research department. Because of the nature of the product and a high level of management enthusiasm for it, the company felt that the product should be put into production as rapidly as possible. The cost estimates used to justify the venture were made without full design information and without a comprehensive market analysis, as evidenced by a comparison of actual with estimated costs; the values actually realized in this example show little relationship to those estimated. In this case nobody cared, because the project turned out to be so successful. The cost engineer's function was simply to come up with some numbers quickly, using whatever short-cut techniques he could muster. Table 13.1 summarizes the results.

As it turned out, the new chemical product was much better than any competitive products on the market at the time, and the first-year sales exceeded the estimates by 20 percent. This positive variance, coupled with a fixed investment that was nearly 30 percent below the estimate, resulted in a percentage return of 29 percent, almost double the estimated return of 15 percent.

This example is unusual in that everything turned out better than anticipated. Since a new venture must show a reasonably attractive return in the preinvestment analysis before it is approved, cost estimating errors on the high side (giving lower profits and percentage returns) are not usually discovered; because the return is so low, the projects are never undertaken. Consequently, the postinvestment analysis usually reveals projects that appeared to be economically attractive in the initial analysis, but performed less well than expected. If this example carries a message, it is to estimate the lowest acceptable return and then try to exceed it. Such "games," and many others, are privately known to experienced managers. The adoption of those games in engineering economics can defeat its primary purpose—to identify the most attractive project(s) among a group of many.

Table 13.1   Inaccurate Estimates but a Financial Success—Facilities for a New Chemical Product

	Estimated ($)	Actual ($)	Ratio of Actual to Estimated
Total capital requirement	1,500,000	1,057,000	0.71
Net sales	1,200,000	1,444,000	1.20
Costs			
Production costs	300,000	361,000	1.20
General overhead	210,000	289,000	1.38
Fixed charges	240,000	179,000	0.75
Total costs	750,000	829,000	1.11
Net profit before taxes	450,000	615,000	1.37
Income taxes	225,000	302,000	1.34
Net profit after taxes	225,000	313,000	1.39
Annual return on capital	15.0%	29.6%	1.97

Example 2:   Manufacturing Facilities for a Plastics Intermediate

This project was proposed because of the patent protection available on certain manufacturing processes. The preinvestment analysis indicated a high return on investment and a rapid payback. Table 13.2 shows the estimated and actual figures for the plant's first year of operation.

The first investment was slightly underestimated, but the chief problem here was failure to anticipate the market and the competition properly. The initial reaction of competitors was to cut prices and increase their promotional activities, resulting in lower sales at lower prices than estimated for this new plant. The net sales actually realized were more than 40 percent below the estimated first-year sales, resulting in a percentage return of only a sixth of that projected. Other variations between actual and estimated results were of a minor nature, attributable mainly to the lower scale of operation.

This example emphasizes the need for thorough market research. Even a good product in a strong market will be successful only if the company is capable of meeting its competition in terms of price, distribution, and service, and these factors should all be considered before committing capital to a new venture. A poor estimate of the effect of competitor's moves on the sales volume of a new product can result in a financial disaster.

Table 13.2   Inaccurate Estimates of Competition Meant Financial Failure for Facilities for a Plastics Intermediate

	Estimated ($)	Actual ($)	Ratio of Actual to Estimated
Total capital requirement	815,000	842,000	1.03
Net sales	1,150,000	682,000	0.59
Costs			
Production costs	650,000	416,000	0.64
General overhead	110,000	96,000	0.87
Fixed charges	115,000	126,000	1.10
Total costs	875,000	638,000	0.73
Net profit before taxes	275,000	44,000	0.16
Income taxes	140,000	21,000	0.15
Net profit after taxes	135,000	23,000	0.17
Annual return on capital	16.6%	2.7%	0.16

EXAMPLE 3:   Acquisition of a Metal Fabricator

The experience report shown in Table 13.3 describes the results of an acquisition. The company whose assets were purchased was a leading customer of the firm making the acquisition, and the primary purpose of the transaction was to protect a position as a raw materials supplier. This objective was attained, even though only a modest return on invested capital was realized. Thus, the venture could be considered a successful acquisition; the company operated satisfactorily and performed generally as expected.

Net income was substantially less than estimated (chiefly as a result of price declines), yet the total profits realized on the sale of raw materials to the fabricating operation made up for most of the deficit. In terms of the company's overall operations, the net profit was just 13 percent less than expected, and the IRR was off by about the same amount.

Some of the major problems encountered in studying prospective acquisitions include difficulties involved in making good estimates of operating costs, in realistically evaluating a company's market strengths and weaknesses, and in properly assessing the technical and managerial competence

Table 13.3   Experience Report on Acquisition of a Metal Fabricator

	Estimated ($)	Actual ($)	Ratio of Actual to Estimated
Total capital requirement	800,000	792,000	0.99
Net sales	1,000,000	1,144,000	1.14
Costs			
Production costs	740,000	1,019,000	1.38
General overhead	85,000	134,000	1.58
Fixed charges	120,000	128,000	1.07
Total costs	945,000	1,281,000	1.36
Net profit on manufacturing operations	55,000	(137,000)	—
Profit on sales to manufacturing	85,000	245,000	2.88
Net profit before taxes	140,000	108,000	0.77
Income taxes	65,000	43,000	0.66
Net profit after taxes	75,000	65,000	0.87
IRR	9.4%	8.2%	0.87

of people in the acquired company. These difficulties are often compounded by the reluctance of the company to reveal detailed operating data or any adverse information during the negotiations; only the good side is presented to the prospective buyer.

SUMMARY

Economic justification of capital expenditures actually involves two phases: (1) the investment evaluation, conducted prior to any capital commitment, and (2) the performance review, made after the project is in operation. The adequacy of investment analysis techniques can be measured only in terms of how well they anticipated actual investment performance. The performance review helps in improving estimating and project evaluation techniques, as well as in identifying problem areas and providing management with useful information regarding the progress of new ventures. The key considerations in conducting an effective postinvestment appraisal are: (1) assigning the right people to the job, (2) selecting the appropriate projects for review, (3) conducting the appraisal at the proper time, and (4) reporting the results of the audit in a suitable format for interpretation. In most cases, actual results are less favorable than estimated, since estimates indicating poor results discourage undertaking the project and the poor estimates are never revealed. Although it is important to guard against overoptimism in investment evaluation, the economic consequences of overlooking good projects are equally real, but less identifiable, than losses suffered from entering unprofitable ventures. The experience reports of companies in various fields emphasize the need for, and illustrate the consequences of not having, thorough cost and market research coupled with sound estimating techniques in the evaluation of new ventures. Properly administered, postinvestment appraisals can lead to a better understanding of what has happened in the past, can improve the engineer's capabilities for future performance, and can even provide material of value in developing new clients.

14

Breakeven Analysis

Before you make a profit, you have to make a sale.

Author Unknown

Breakeven analysis is a useful tool in business and profit planning, since many management decisions require that the effect of changing sales volumes, markups, and overheads be viewed in terms of the company's overall profit position. One of the major problems encountered by management in making these types of decisions is being able to visualize correctly the interactions among all the different elements that must be considered.

Breakeven analysis can clearly illustrate the essential relationships among such factors as direct job costs, overhead costs, percentage markups, sales volumes, and profits. By clarifying these relationships, management can objectively evaluate the impact of changes in the different elements and can select appropriate strategies for working toward the firm's objectives.

BREAKEVEN TERMINOLOGY

In breakeven analysis, costs are broken down into their fixed, semivariable, and variable elements.

Fixed costs remain constant (theoretically) over a period of time, regardless of the quantity of work performed by the firm. Fixed costs include many of the general overhead items.

Semivariable costs may actually be fixed within certain relatively narrow ranges of sales volumes, but generally increase in some relation to sales. Semivariable costs include some capacity-related general overhead items, plus any additional administrative expenses necessary to handle heavy work loads. If semivariable costs vary directly with the volume of work, they can be included with variable (or direct) costs; if they are stable over well-defined ranges, they may be handled in the same manner as fixed costs.

Variable costs, also referred to as *direct costs, job costs,* or *out-of-pocket costs* are those that vary directly with the amount of work undertaken. Job-related overheads may be included in this category.

Total income and *total sales* are used synonymously, referring to the total amount of money generated by the business or project being considered.

Total costs are made up of the sum of all fixed, semivariable, and variable costs incurred over a specified time period.

Profit refers to the difference between total income and total cost whenever the income is greater than the cost. Unless otherwise specified, profit means profit before income taxes are paid.

Loss refers to the difference between total cost and total income whenever the cost is greater than the income.

The breakeven point is the sales volume at which there is neither profit nor loss, or where total income is equal to total cost.

The breakeven point may also be expressed as a percent of capacity, or in terms of units of production, instead of as a dollar sales volume.

The *markup* is a percentage added to estimated direct costs (usually direct labor and materials). The markup must be sufficient to recover all fixed costs. Ideally, the markup is also sufficient to return a profit on the invested time and capital.

The variable cost rate is the ratio of direct (variable) costs to volume and is expressed as a decimal:

$$R = \frac{D}{V}$$

INFORMATION REQUIREMENTS

Cost data provide the basis for breakeven analysis. These basic cost data should be compiled, analyzed, and related to different levels of sales volume.

Historical accounting records usually provide the most convenient sources of information used in breakeven analysis, although, in the absence of such records, good estimates may be satisfactory for planning purposes.

Table 14.1 shows a summary operating statement (or profit-and-loss statement) for a typical manufacturing company with annual sales of $1,000,000. By using only these basic data, a useful and meaningful breakeven analysis can be made.

USE COSTS AND SALES TO CALCULATE BREAKEVEN POINT

From the fixed costs of $230,000 shown in the operating statement, and the direct job costs figure of $670,000 representing all variable costs, this firm's break-even point can be determined.

The breakeven point, by definition, is that level of sales at which total income equals total cost. Total cost, in turn, includes both fixed and variable costs. To find the breakeven point for this firm, the variable costs, which when added to the $230,000 fixed costs equal total income at that point, must be calculated. In making this calculation, the total variable cost of $670,000 is assumed to have been accumulated at a constant rate over the $1 million sales volume; therefore, for each dollar of sales, $670,000/$1,-000,000, or $0.67, in variable costs are incurred. Stated another way, variable costs are equal to 67 percent of the sales volume.

Table 14.1   Breakeven Points Can Be Calculated from an Operating Statement for a Manufacturing Company

Total sales		$1,000,000
Cost of operations		
Direct job costs	$670,000	
Fixed costs	230,000	
Total costs		900,000
Net profit before income taxes		$ 100,000

Having accumulated the necessary information, the basic formula $V = F + D$ can be used to calculate the breakeven point; V equals the volume of sales required to break even, F is the total amount of fixed cost, and D represents the total amount of direct costs at the breakeven point, expressed as a percentage of sales volume. In this case, $0.67V$ can be substituted for D. Substituting the data from the preceding operating statement into the breakeven formula then gives:

$$\$230,000 + 0.67V = V$$

$$0.33V = \$230,000$$

$$V = \$697,000$$

The breakeven point for this firm, then, occurs at a sales volume of $697,000. Should the company fail to achieve this volume, a loss would result; and the firm begins to realize a profit only after a sales volume of $697,000 has been exceeded. The breakeven calculation can be checked by working backward from the breakeven point, as shown in Table 14.2.

INDUSTRY BREAKEVEN POINTS REFLECT COST STRUCTURES

In the preceding chapter, typical cost structures were presented for several major industry groups. By using the fixed-variable cost relationships shown for these industries, some interesting comparisons can be drawn regarding their relative breakeven points.

When the variable cost rate for a firm is known (0.67 for the manufacturing firm in the preceding example) and the fixed costs can be identified for a specified sales volume, the basic breakeven formula $V = F + D$ can be restated.[*]

$$V = \frac{F}{1 - R}$$

[*] $V = F + D$ and $RV = D$;
$V = F + RV$;
$V - RV = F$;
$V(1 - R) = F$;

$$V = \frac{F}{\sqrt{(1 - R)}}$$

Table 14.2   Breakeven Volume for Company in Table 14.1

Sales volume at breakeven point	$697,000
Less variable expenses at 67% of sales	467,000
Amount remaining to cover fixed costs	$230,000
Less fixed costs	230,000
Net profit or loss	0

Again, V and F represent the breakeven volume and fixed costs, and R is the variable cost rate. This formula can be applied to the industry cost data to identify the breakeven points for each industry, as shown in Table 14.3. An assumption was made that in all 14 of the industries a typical firm could realize a 10-percent pretax profit margin on a $1 million sales volume. The breakeven points for these industries range from a low of $565,000 for food stores, with their low fixed costs and high-variable-cost rate, to a high of $853,000 for public utilities, with just the opposite type of cost structure.

Some idea of the relative markups applied to direct costs in these industries can also be obtained from the data in Table 14.3. It is apparent that the high-fixed-cost industries must apply higher percentage markups on their direct or out-of-pocket costs to recover their fixed costs. The average percentage markup (M) on direct costs is

$$M = \frac{100}{R} - 100^{\circ}$$

where R is the variable cost rate expressed as a decimal, as in Table 14.3. For the 14 industries shown, markups range from a high of 213 percent for public utilities $[(100/0.32) - 100]$, where the direct costs of energy production are far outweighed by the fixed costs of production and distribution, to a low of 30 percent for food stores $[(100/0.77) - 100]$, where the pur-

---

$^{\circ}$ $(100 + M)D = 100V;$
   $100D + MD = 100V;$

$$M = \frac{(100V - 100D)}{D} = \left(\frac{100V}{D}\right) - 100.$$

Since $D = RV,$

$$M = \left(\frac{100V}{RV}\right) - 100 = \left(\frac{100}{R}\right) - 100.$$

Table 14.3   Breakeven Point for Typical Firms in Selected Major Industry Groups[a]

Industry	Fixed Costs ($1000)	Variable Cost Rate (R)	Markup (%)	Break-even Volume ($1000)
Agriculture	230	0.67	49	697
Mining	320	0.58	72	762
Construction	220	0.68	47	687
Manufacturing	230	0.67	49	697
Railroad transportation	500	0.40	150	833
Motor freight transportation	190	0.71	41	655
Air transportation	310	0.59	70	756
Telephone and telegraph	510	0.39	156	836
Public utilities	580	0.32	213	853
Wholesale trade	160	0.74	35	615
Retail trade	190	0.71	41	655
Food stores	130	0.77	30	565
Business services	230	0.67	49	697
Engineering and architectural services	160	0.74	35	615

[a] Assumes a 10% pretax profit margin at $1,000,000 sales volume.

chased cost of goods sold represents a high proportion of the retail selling price.

THE BREAKEVEN CHART

Although the arithmetic calculations involved in breakeven analysis are simple and should be thoroughly understood, the use of a breakeven chart provides a more vivid picture of exactly where the firm stands with respect to its breakeven point. Also, the relationships among the various factors that influence the breakeven point can best be expressed graphically. The breakeven chart, then, offers a convenient means of illustrating several important cost relationships that would otherwise be difficult to express. Many feasibility reports could probably benefit substantially by substituting a few well-chosen charts for a few thousand words.

The same information is required to construct a breakeven chart as is used in computing the breakeven point arithmetically. The conventional breakeven chart assumes that fixed costs remain constant, regardless of the sales volume, and that variable or direct job costs change in direct proportion to sales.

Figure 14.1 shows the essential features of a breakeven chart for the manufacturing firm described in Table 14.1. The vertical axis represents both income and cost, and the horizontal axis shows total sales volume. The units of measurement on both scales are dollars.

At zero sales, income is also zero. Therefore, the total income line is a straight line passing through the origin, representing the income increasing as sales increase. Since the company's income is presumably derived solely from sales, the total income shown on the vertical axis is aways equal to the total sales volume shown on the horizontal axis. Next, the total amount of fixed costs is plotted on the graph. Fixed costs in this case total $230,000 throughout the range of sales shown on the breakeven chart. Variable costs are then added to fixed costs to arrive at total costs, with the slope of the variable cost line depending on the rate at which these costs are incurred—

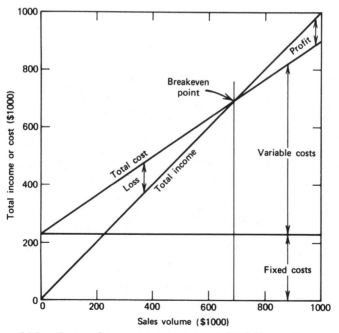

Figure 14.1. *Cost and income intersect to establish the breakeven point.*

in this case, at 67.0 percent of sales. These variable costs are plotted above the fixed cost line, thereby giving the total cost of operations for any given level of sales. The total cost line—the sum of fixed and variable costs—indicates the total cumulative amount spent at a specific sales volume.

The point at which the total cost line intersects the total income line—$697,000 in this example—is the breakeven point. To the left of this point, the vertical distance between the total income and the total cost lines indicates a net loss; to the right, it represents the net profit. At the $697,000 breakeven point, sales revenues are exactly matched by the two elements of cost: $230,000 in fixed charges and $467,000 in variable costs. At a sales volume of $800,000, profits of $34,000 result; with sales of only $500,000, a net loss of $65,000 accrues.

The breakeven point can be described in terms of the firm's operating capacity as well as sales volume, should this measure be more meaningful to management. For example, if the firm's annual capacity were $1,500,000, the breakeven point would occur at about 46.5 percent of capacity. To earn net profits of $34,000 would require operating at 53.3 percent of capacity. In either case, the procedure is the same, differing only in the units of measurement along the horizontal axis of the graph.

FIXED COSTS ARE NOT ALWAYS FIXED

In the preceding example, fixed costs were assumed to remain constant regardless of the level of sales. In most cases, however, fixed costs do not exactly merit their name. They are apt to increase with the firm's capacity to do business. Usually, these additional overhead costs are not incurred gradually but come in steps, with each step covering a sizable range—just as adding another generator set to a power plant increases plant cost in a single step, which then remains at that level over the next range of generating capacity. Figure 14.2 illustrates the effect of varying fixed costs on a firm's breakeven point. The scales are the same as in the previous example. As indicated in Figure 14.2, the effect of lowering the level of fixed costs is to cause the total cost line to intersect the total income line more quickly, indicating a lower breakeven point. Lowering the fixed cost from $230,000 to $150,000 reduces the breakeven point from $697,000 to $455,000; thus, an $80,000 reduction in fixed costs results in a $242,000 reduction in the breakeven point.

By reducing fixed costs by any given percentage, the breakeven point is

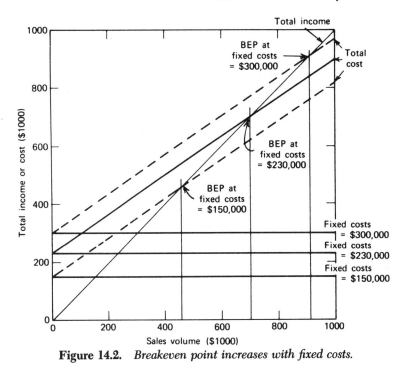

Figure 14.2.   *Breakeven point increases with fixed costs.*

reduced by that same percentage. Or by increasing the fixed costs by any given percentage, the breakeven point is increased by a like percentage.

Hence, increasing fixed charges from $230,000 to $300,000 moves the breakeven point farther away, from $697,000 to $909,000. This means that $212,000 in additional sales has to be generated just to make up for the additional $70,000 in fixed cost obligations. Again, the same percentage increase must be made in sales as in fixed costs to retain the same position of economic equilibrium.

Figure 14.2 resulted from the assumption that fixed costs were independent of sales volume. More than likely, however, a given level of sales will require a similarly given fixed cost—for example, equipment and the number of salespersons and sales supervisors. That sales force can be efficient over a fairly wide range of volume, but the success of their product will eventually require a larger investment in equipment and more sales service personnel. In the process, fixed costs rise to a new plateau. Figure 14.3 illustrates such a pattern. That figure shows how the same company's operations might look if these three levels of fixed costs—$150,000, $230,000, and

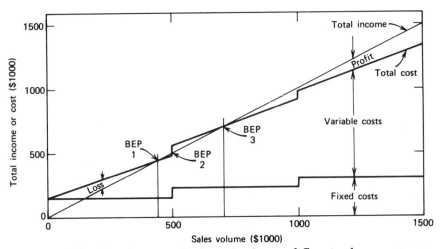

Figure 14.3. *Effect of varying fixed cost levels over different volume ranges on the breakeven point (BEP).*

$300,000—are incurred over specific ranges of sales volume. In this figure, fixed costs of $150,000 are associated with sales volumes of zero to $500,000; $230,000 is incurred at sales volumes between $500,000 and $1,000,000; and $300,000 in fixed charges is necessary to support sales between $1,000,000 and $1,500,000.

The first breakeven point for this operation occurs at sales of $455,000. Profits rise to $15,000 at a $500,000 sales volume, at which time additional fixed charges must be incurred to support any additional volume, and the breakeven point rises to $697,000. Breaking even at the higher overhead requires $242,000 more sales, and to make the same amount of profit as was made at the $500,000 volume ($15,000) now requires sales of $742,000— also $242,000 more than before.

The breakeven chart provides a useful visual representation or model of a firm's overall operating picture, showing important relationships between income and costs. It vividly emphasizes the sometimes alarming effect of overhead or fixed costs on the breakeven point. The effect of overhead costs on the breakeven point depends primarily upon the rate at which variable costs are incurred; in some cases, each change of $1 in fixed costs may change the breakeven point by $20 or more. Failure to consider the effect of fixed costs in a firm's pricing policies can, and often does, result in making an apparent profit on every job, but still losing money in total.

MARKUPS—THE ULTIMATE CONTROL OVER BREAKEVEN POINT

Whether selling new construction, engineering services, or retail items, the seller's markup—the percentage he adds to his direct costs to cover overhead and profit—has an important effect on the volume of sales he must attain to break even. The higher the markup, the lower the breakeven point; the lower the markup, the higher the breakeven point.

Markup is indirectly related to discounts. Books, as a familiar example, usually carry about a 40-percent discount to the retailer; in other words, the selling price of a $10 book is made up of the bookseller's relatively high direct cost ($10 less 40%, or $6) and his $4 markup. From the bookseller's standpoint, this $4 is a 66.7 percent markup on his out-of-pocket direct cost ($1.667 \times \$6 = \$10$). On this basis, his direct costs—that is, the actual out-of-pocket costs of the books he sells—are accruing at a rate of 60 percent of sales. His problem is to sell enough books to cover his overhead and return a profit, from the $4 markup on each book.

The need for sufficient markup to cover overhead and make a profit is also typical of engineers, contractors, and businessmen in general.

Not all businesses uses the term markup. For example, a consulting engineer or other type of service firm (legal, accounting, advertising, etc.) might typically price its services by applying a multiplier of from 2.0 to 3.0; these multipliers actually represent markups (on direct costs) of 100 and 200 percent.

Markups can be used to calculate direct costs as a percent of sales. At the 100 percent markup (or 2.0 multiplier), the firm's direct costs are accruing at a rate of 50 percent of sales. At the 200 percent markup, the rate at which variable costs are incurred is just 33.3 percent of sales. Similarly, a general contractor's 10-percent markup on direct, or variable, costs translates to variable costs accruing at 90.9 percent of sales.

A few more fortunate contractors may be able to charge a 25-percent markup on their direct costs; their variable cost rate, then, is 80 percent.

The relationship between the percentage markup and the variable cost rate is direct. A low variable cost rate is associated with a high markup and a low breakeven point, and a high variable cost rate with a low markup and a high breakeven point.

Figure 14.4 illustrates how and why the breakeven point varies with markup. In this example, the firm has annual fixed costs of a relatively low

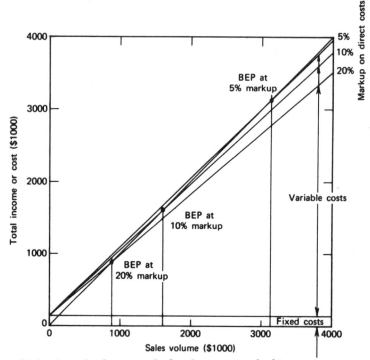

Figure 14.4.   *Low fixed costs make breakeven point highly sensitive to markup.*

$150,000. With a 10-percent markup, the breakeven point occurs at a sales volume of $1,650,000. This volume is matched by total costs of the same amount, made up of $1,500,000 in direct (or variable) costs plus the 10-percent markup that covers the $150,000 in fixed costs. The variable cost line has a slope of $909,000 in $1,000,000, or 90.9 percent of sales.

A higher markup decreases the slope of the variable cost line, causing it to intersect the total cost line more quickly. A 20-percent markup results in a breakeven point of $900,000, representing a slope of $833,000 in $1,000,-000, or 83.3 percent of sales. Similarly, lowering the markup, by raising the slope of the variable cost line, causes it to intersect later with the total cost line, denoting a higher breakeven point. For a 5-percent markup, the breakeven point is at $3,150,000—made up of $3,000,000 in direct costs and $150,000 in fixed costs. Variable costs in this case are incurred at a rate of 95.2 percent of sales.

Such is the fate of the contractor with his usually low fixed costs. The bookseller, however, with his 66.7 percent markup, is able to recover his relatively low $50,000 in overhead on a sales volume of only $125,000. The consultant, with a 200-percent markup, can break even with just $75,000 worth of business.

BREAKEVEN CAN BE CALCULATED FROM MARKUP

The relationship among fixed charges, percentage markups on direct job costs, and the volume of work required to break even, can be expressed as:

$$\frac{V}{F} = \frac{100 + M}{M} = \frac{100}{M} + 1$$

where V is the breakeven volume, F is the fixed charges or overhead, and M is the percentage markup on direct costs, expressed as a percent.

Using this formula, a 10-percent markup requires a sales-to-overhead ratio of 110/10, or 11.0. Similarly, a 20-percent markup requires a sales volume of 120/20, 6.0 times the overhead. A 100-percent markup results in a breakeven volume of 200/100 or 2.0 times the overhead. The relationship between the percentage markup on direct costs and the ratio of sales to fixed costs is shown graphically in Figure 14.5.

SIGNIFICANCE OF RELATIONSHIPS

Since a firm's breakeven point is a direct function of its overhead costs and an inverse function of percentage markups, any change in either of these factors is greatly amplified in the amount of work the firm must obtain to break even. Working on a 10-percent markup, each $1 of overhead must be matched by $11 in sales. Thus, if an additional $1000 of overhead expense is taken on, an additional sales volume of $11,000 must be generated just to maintain the same profit position. Reducing overheads by a like amount reduces the breakeven point by the same $11,000.

Important changes in a firm's financial performance can be brought about by varying markups and sales volumes. In highly competitive industries such as construction, it may not be possible to raise markups, but equally beneficial results may be obtained by reducing overheads. Firms engaged in selling unique services or products may, however, find it easier to maintain profits by raising prices.

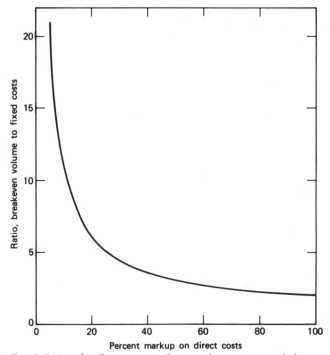

Figure 14.5.   *Relationship between markup on direct costs and the ratio of sales to fixed costs.*

SUMMARY

Breakeven analysis is a useful tool in defining and describing the relationships between a firm's sales revenues, fixed and variable costs, markups, and profits. A company's breakeven point can be calculated from financial data usually available in its financial statements and varies widely according to the characteristics of the industry in which it operates. High-fixed-cost operations have relatively high breakeven points, whereas industries and companies operating with a high variable-cost rate have relatively low breakeven points. A company's breakeven point can be calculated from the formula $V = F/(1 - R)$, where V is the breakeven volume, F is the fixed cost, and R is the variable-cost rate, expressed as a decimal. In addition to the variable-cost rate, fixed costs and percentage markups on variable costs are the primary determinants of a firm's breakeven point. A given percentage change in fixed costs changes the breakeven point by the same percentage and in the same direction. The mathematical relationship between fixed costs (F), percentage markups on variable costs (M), and the break-even volume (V) can be expressed as $V/F = (100 + M)/M$. Thus, a 10-percent markup requires sales of 11 times the fixed costs to break even, and a 20-percent markup lowers the sales/fixed cost ratio to 6.0. Overhead cost control, therefore, is particularly important in highly competitive industries such as construction and manufacturing, since each additional dollar of fixed cost incurred may be multiplied many times in the additional amount of sales income that must be generated to break even. A detailed study of the uses of breakeven analysis can be found in *Profit Planning Decisions with the Breakeven System* by Spencer A. Tucker (Thomond Press, New York, 1980).

15

Economic and Financial Models

Reduce the problem to as simple a form as possible, then find the answer by the most direct means.

Author Unknown

Most engineers are familiar with and experienced in the use of mathematical models in connection with various types of design problems. These mathematical models, whether dealing with the deflection of a column or beam, with the flow of a fluid through a pipe, or with the transmission of electrical energy, are extremely useful—in fact, essential—in presenting an abstract and simplified picture of a realistic physical process.

An engineering model expresses, in a system of mathematical equations, the relationships between several variables and constants. Some variables are self-explanatory, some must be explained; some are independent variables, others are dependent variables; some are endogenous variables, others are exogenous variables. Likewise, some of the constants are peculiar to the materials used, whereas others reflect the external design conditions.

Mathematical models are indispensable to engineering studies because they give a simplified picture of what is likely to take place and focus attention on the most critical aspects of the situation. Economic models are equally useful in feasibility studies.

VARIATIONS CHARACTERIZE ECONOMIC MODELS

Economic models define the relationships between several economic variables, expressing these relationships in equations that usually include both variables and constants, or parameters. Since economic conditions are much less predictable than conditions controlled by the laws of physics, chemistry, and mathematics, economic models often require the examination of a wide range of possibilities. Sometimes, there are so many different possibilities that the problem is dismissed as being impossible to solve—a most unsatisfactory conclusion. In many cases, however, the computer is a way of approaching a complex economic problem. The computer can perform essentially the same operations in the same sequence as would otherwise be handled manually.

The most valuable use of the computerized economic model is not just in turning out answers, but in showing how much difference a change in the assumptions or variables makes in the answer. The actual computer time required to examine different assumptions and recalculate an entire problem is almost insignificant. The computer can, in just a few seconds, perform calculations that might otherwise require many man-hours of manual effort.

Economic models can be either convenient or flexible, but are seldom both. A convenient and simple model contains mostly constants and only a few variables. A flexible and realistic model probably has a large number of variables and few, if any, constants.

Since profitability is the ultimate goal of any economic enterprise, the determination of the probability of profit is the ultimate objective of most economic models.

PROFITS REFLECT ATTENTION TO THE CRITICAL FACTORS

The factors that ultimately determine the profits to be made from a business venture and the relationships among those factors are illustrated in Figure 15.1.

Profit, the target for any business venture, is directly determined by three factors: cost, price, and volume. Each of these three factors is, in turn,

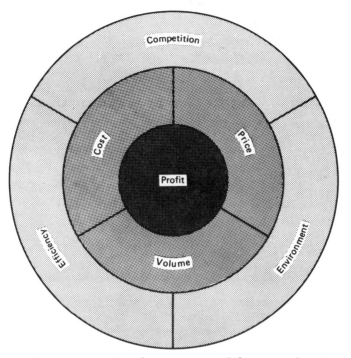

Figure 15.1. *Profits reflect three primary and three secondary factors.*

influenced by other considerations that may be beyond the company's direct control. Whether or not they can be controlled, they must be recognized, understood, and acted upon (or reacted to).

The economic environment determines how large the total market for a specific type of product or service is; competition determines how much of that total market can be obtained at a given price, quality, or service level; and the company's efficiency determines how much it costs to produce and market the product or service at the required scale of operations. A profit orientation, therefore, requires that operations be planned on the basis of the business climate and outlook, priced on the basis of competitive factors, and carried out as efficiently and economically as possible.

THE FACTORS COMBINE TO YIELD RETURN ON INVESTMENT

Price, cost, and volume are the fundamental determinants of a venture's profits. Profits, related to investment, define the venture's profitability.

Several important factors enter into the determination of price, cost, and volume. Some of these factors are shown in Figure 15.2, where interrelationships between balance sheet and operating statement accounts are illustrated.

The balance sheet items—primarily working capital and fixed investment—constitute the total investment applicable to a given project. Labor, materials, and variable overheads (from the operating statement) make up

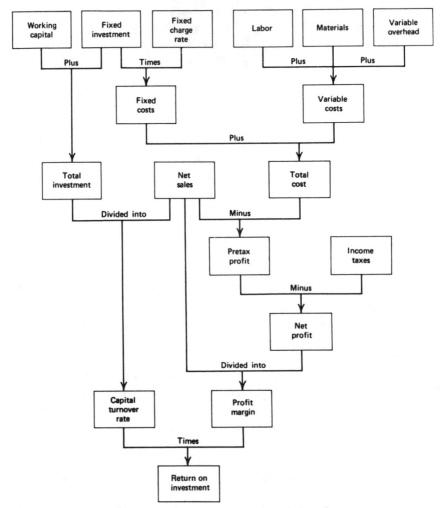

Figure 15.2. *Relationships between factors included in the economic model.*

the bulk of variable or direct production costs. By applying an appropriate factor (representing capital charges) to fixed investment, the annual fixed costs are established and, added to variable costs, give the total cost of doing business under the prescribed conditions. Net sales less total costs give the pretax profit margin, which can be reduced to net profit by deducting the applicable income taxes.

The resulting investment, sales, and profit figures can then be combined to form several significant financial ratios. Sales divided by investment gives the rate of capital turnover; profit divided by sales gives the profit margin. Finally, the profit margin multiplied by the capital turnover rate gives the return on invested capital (ROI):

$$\frac{\text{profit}}{\text{sales}} \times \frac{\text{sales}}{\text{investment}} = \frac{\text{profit}}{\text{investment}}$$

Submodels provide another means of analysis. A profitability model can be broken down into a group of separate and distinct submodels. For example, a complete profitability model might include six separate stages or submodels: a market model, a cost model, an investment model, a depreciation model, a profit model, and a cash flow model. Table 15.1 summarizes the essential features of the various submodels. The market, cost, and investment models require considerable work and are ideal candidates for using the "what if" capabilities of electronic spread sheets. The other three models simply define the computational procedures to be applied. Finally, the cash flow model is used to develop the appropriate measures of profitability or project desirability. It also will be more informative if assembled on an electronic spread sheet.

THE MARKET MODEL

The market model is concerned primarily with the price-volume relationship: how much can be sold at a given price, or what price is required to attain a specific volume. For most products and services, volume increases as the price decreases, and increasing prices are usually accompanied by declining sales. The market model allows the engineer economist to estimate the net income to the company over a period of years, based on various combinations of prices and sales volumes. Price is usually considered the independent variable, and volume the dependent variable; total income

Table 15.1   Major Elements of Financial Submodels

Income submodel
    Units sold
  × Net price per unit
  = Net sales
  − Selling costs
  = Net income from sales
Manufacturing cost submodel
    Units produced
  × Unit cost (material + labor + variable overhead)
  = Direct manufacturing cost
  + Fixed overhead
  + Capital charges
  = Total manufacturing costs
Investment submodel
    Depreciable investment (buildings and equipment)
  + Expensed and amortized investment (engineering, research and development, startup)
  + Land
  + Working capital (receivable and inventory)
  = Total capital requirement
Depreciation submodel
    Total depreciable investment
  × Depreciation rate
  = Annual depreciation charge
Income summary submodel
    Net income from sales
  − Total manufacturing cost
  = Net operating income
  − Depreciation
  − Expensed investment
  = Net profit before income taxes
  − Income taxes
  = Net profit after taxes
Cash flow submodel
    Net profit after taxes
  + Depreciation
  + Expensed investment
  = Net cash inflow
  − Net cash outflow (total capital requirement)
  = Net cash flow

is the product of price and volume less direct selling and promotional costs. In analyzing many types of new projects, sales volume can be set at some percent of plant capacity (so many tons per year, rental units available, seating capacity, etc.), and the prices needed to attain that degree of utilization can then be determined, based on the market situation. Schedule A (Table 15.2) shows an example of a sales and income forecast covering a new project over its economic life. In this example, the plant will be completed in 2 years, followed by 8 years of production. The product's selling price, initially $1000 per unit, will decrease to $975 per unit when the production capacity of 6000 annual units is reached. Selling costs are estimated to be 10.0 percent of net sales.

THE COST MODEL

The cost-volume relationship is just the opposite of the price-volume relationship. Here, as volume increases, cost decreases; and higher costs are usually associated with lower volumes. The cost model considers direct labor and materials, variable and fixed overheads, and capital charges—all independent variables—to arrive at the dependent variable, the manufacturing cost as a function of volume. If a nonmanufacturing project is involved, only the terminology will be different; instead of manufacturing or production costs, for example, the term operating costs would be more descriptive. Schedule B (Table 15.3) shows the manufacturing costs for the

Table 15.2   Schedule A: Sales and Income Forecast

Year	Units Sold	Net Price Per Unit	Net Sales	Selling Costs	Net Income from Sales Per Unit	Annual
0	0					0
1	0					0
2	4,000	1,000	4,000,000	400,000	900.00	3,600,000
3	4,500	1,000	4,500,000	450,000	900.00	4,050,000
4	5,000	1,000	5,000,000	500,000	900.00	4,500,000
5	5,500	1,000	5,500,000	550,000	900.00	4,950,000
6	6,000	975	5,850,000	585,000	877.50	5,265,000
7	6,000	975	5,850,000	585,000	877.50	5,265,000
8	6,000	975	5,850,000	585,000	877.50	5,265,000
9	6,000	975	5,850,000	585,000	877.50	5,265,000

Table 15.3 Schedule B: Manufacturing Costs

Year	Units Produced	Units Costs ($) Materials	Labor	Variable Overhead	Total	Period Costs ($) Fixed Overhead	Capital Charges	Total	Total Manufacturing Cost ($) Per Unit	Annual
0	0									
1	0									
2	4,000	400.00	60.00	60.00	520.00	50,000	250,000	300,000	595.00	2,380,000
3	4,500	400.00	55.00	60.00	515.00	50,000	250,000	300,000	581.67	2,618,000
4	5,000	400.00	50.00	60.00	510.00	50,000	250,000	300,000	570.00	2,850,000
5	5,500	400.00	45.00	60.00	505.00	50,000	250,000	300,000	559.55	3,078,000
6	6,000	400.00	40.00	60.00	500.00	50,000	250,000	300,000	550.00	3,300,000
7	6,000	400.00	40.00	60.00	500.00	50,000	250,000	300,000	550.00	3,300,000
8	6,000	400.00	40.00	60.00	500.00	50,000	250,000	300,000	550.00	3,300,000
9	6,000	400.00	40.00	60.00	500.00	50,000	250,000	300,000	550.00	3,300,000

project whose sales were described in schedule A. Here, materials costs and variable overheads are assumed to remain constant on a per-unit basis, while labor costs decrease as production volume increases. The period costs (or time-related charges) stay at $300,000 per year throughout the project's operating life.

THE INVESTMENT MODEL

The magnitude and timing of capital requirements require some careful analytical work, as well as some intuitive judgment. Here, a sequence network with a critical path proves invaluable in planning the investment schedule. Included in the capital requirements are depreciable investments—consisting primarily of buildings and equipment; expensed or amortized investments, such as research and development, engineering, and startup costs; and nondepreciable investments, such as land and working capital, which consists mainly of the funds needed to carry accounts receivable and inventories. The investment model's output is the total capital required by years over the project's economic life. Schedule C (Table 15.4) shows the capital requirements by years for the sample project. Year 0 represents the preliminary design and land acquisition phase; year 1 includes most of the actual plant construction, along with additional engineering work; and year 2 sees completion of the plant, followed by startup and initial production. Working capital requirements in the example are estimated at 10.0 percent of total manufacturing cost for inventories and at 10.0 percent of net sales for accounts receivable. The capital requirements for years 3–9 represent only the *additional* working capital required each year to keep up with increases in sales and manufacturing costs.

THE DEPRECIATION MODEL

The depreciation model should be set up to handle depreciable assets having different recovery periods, computing the annual recovery allowance and the remaining unrecovered balance on an annual basis. The recovery methods that allow an accelerated write-off are favored by most large firms. Appropriate allowances must also be made for investment tax credits when applicable. Output from the depreciation model consists of total annual recovery deductions. Schedule D (Table 15.5) shows the recovery charges for

Table 15.4 Schedule C: Capital Requirements

Year	Recoverable Investment Buildings	Equipment	Expensed Investment Engineering R & D	Startup	Land	Other Capital Working Capital Receivables	Inventory	Total Capital Requirements
0	0	0	100,000	0	20,000	0	0	120,000
1	500,000	1,200,000	150,000	0		0	0	1,850,000
2	300,000	800,000	50,000	150,000		400,000	238,000	1,938,000
3						450,000	262,000	74,000
4						500,000	285,000	73,000
5						550,000	308,000	73,000
6						585,000	330,000	57,000
7						585,000	330,000	0
8						585,000	330,000	0
9						585,000	330,000	0
Total	800,000	2,000,000	300,000	150,000	20,000	585,000	330,000	4,185,000

Table 15.5   Schedule D: Depreciation

Year	Building (15-Year Recovery) Balance	Building (15-Year Recovery) Amount	Equipment (5-Year Recovery) Balance	Equipment (5-Year Recovery) Amount	Total Balance	Total Amount
0	0	0	0	0	0	0
1	0	0	0	0	0	0
2	800,000	56,000	2,000,000	400,000	2,800,000	456,000
3	744,000	96,000	1,600,000	640,000	2,344,000	736,000
4	648,000	96,000	960,000	480,000	1,608,000	576,000
5	552,000	88,000	480,000	320,000	1,032,000	408,000
6	464,000	80,000	160,000	160,000	624,000	240,000
7	384,000	72,000	0	0	384,000	72,000
8	312,000	64,000	0	0	312,000	64,000
9	248,000	56,000	0	0	248,000	56,000
10	192,000	0	0	0	192,000	0

the example project. In this example, buildings are written off during a 15-year period, and equipment over 5 years. Recovery deductions are computed according to the accelerated recovery tables for 5 and 15 year properties installed after 1986.

THE CASH FLOW MODEL

Except for income tax considerations, the preceding models provide all the data needed for computing net annual cash flows. By assuming an income tax rate to be applied to net earnings (about 50 percent, plus whatever tax surcharges may apply, is typical for large corporations; it is less for small firms), the net annual cash flow, in and out, can be computed over the specified time period. Schedules E and F (Tables 15.6 and 15.7) bring together the data from the previous schedules to provide for the computation of net profit after taxes (in schedule E) and cash flow (in schedule F). In schedule E, the net income from sales comes from schedule A, the manufacturing cost from schedule B, depreciation from schedule D, and expensed investment from schedule C. Income taxes are estimated at 50 percent of the net profit before taxes, and all other figures are calculated. It is further assumed that losses generated during the early years of the project can be applied to the taxes on income from other phases of the company's operations.

Table 15.6 Schedule E: Income Statement[a]

Year	Net Income from Sales	Manufacturing Cost	Net Operating Income	Depreciation	Expensed Investment	Net Profit Before Taxes	Income Taxes	Net Profit After Taxes
0	0	0	0	0	100,000	(100,000)	(50,000)	(50,000)
1	0	0	0	0	150,000	(150,000)	(75,000)	(75,000)
2	3,600,000	2,380,000	1,220,000	456,000	200,000	564,000	282,000	282,000
3	4,050,000	2,618,000	1,432,000	736,000	0	696,000	348,000	348,000
4	4,500,000	2,850,000	1,650,000	576,000	0	1,074,000	437,000	537,000
5	4,950,000	3,078,000	1,872,000	408,000	0	1,464,000	732,000	732,000
6	5,265,000	3,300,000	1,965,000	240,000	0	1,725,000	862,500	862,500
7	5,265,000	3,300,000	1,965,000	72,000	0	1,893,000	946,500	946,500
8	5,265,000	3,300,000	1,965,000	64,000	0	1,901,000	950,500	950,500
9	5,265,000	3,300,000	1,965,000	56,000	0	1,909,000	954,500	954,500
10	0	0	0	0	0	0	0	0

[a] All figures are given in dollars.

Table 15.7 Schedule F: Cash Flow[a]

Year	Net Profit After Taxes	Depreciation Plus Expensed Investment	Net Cash Inflow	Net Cash Outflow	Net Cash Flow	Cumulative Cash Flow
0	(50,000)	100,000	50,000	120,000	(70,000)	(70,000)
1	(75,000)	150,000	75,000	1,850,000	(1,775,000)	(1,845,000)
2	282,000	656,000	938,000	1,938,000	(1,000,000)	(2,845,000)
3	348,000	736,000	1,084,000	74,000	1,010,000	(1,835,000)
4	537,000	576,000	1,113,000	73,000	1,040,000	(795,000)
5	732,000	408,000	1,140,000	73,000	1,067,000	272,000
6	862,500	240,000	1,102,500	57,000	1,045,500	1,317,500
7	946,500	72,000	1,018,500	0	1,018,500	2,336,000
8	950,500	64,000	1,014,500	0	1,014,500	3,350,500
9	954,500	56,000	1,010,500	0	1,010,500	4,361,000
10	0	0	1,127,000	0	1,127,000	5,488,000

[a] All figures are given in dollars.

265

In schedule F, cash inflow is calculated by adding recovery charges (depreciation) and expensed investment charges (from schedule E) to the net profit after taxes (also from schedule E). Cash outflows are taken directly from schedule C. This results in a net cash outflow for years 0–2, followed by 8 years of cash inflows. In year 10, the inflow shown represents the return of land, working capital, and the undepreciated balance of the building account upon completion of the project.

Expensed investment is sometimes handled in a different way than shown here. If it is simply treated as an ordinary expense item, it is not included under capital requirements, and is not added back to net profits to determine cash flow. The end result is the same in either case; the approach used depends upon the company's accounting preferences.

THE DECISION MODEL

From the net annual cash flow and the investment schedules, the relative financial attractiveness of the proposed project can be computed by any or all of several methods.

1.   The desired rate of return can be set, and the present value of future cash flows can be computed by discounting at that rate. This approach can determine if the project is worth further evaluation; if the net present value is positive, then the project is earning *at least* the specified rate. In the example, if the cash flows shown in schedule F are discounted at a 10-percent rate, the project will have a present worth of $2.1 million. With a discount rate of 15 percent, the present worth is $1.2 million; and discounted at 20 percent, the present worth is $519,000. The company's cost of capital determines whether or not the project is an attractive one.

2.   The present value of future cash flows can be set, and the corresponding rate of return (or discount rate) can be computed. If the present value of future cash flows is set at zero, the computed return will be the project's internal rate of return, interest rate of return, profitability index, or DCF rate of return. By employing the DCF technique, the rate of return on invested capital associated with the cash flows summarized in schedule F is 26.1 percent. This return is then considered in view of the risks associated with the project, and the investment decision is made accordingly. As pointed out in Chapter 11 (see Table 11.5), a 25-percent return is a typical profitability objective for a venture involving above-average risk. If this

were a high-risk project, the return would probably be considered inadequate. For an average-risk project, a 26.1-percent return is quite attractive.

3.   The payout time (or payback period) can be computed either with or without interest. If the interest rate is set equal to the project's DCF rate of return, the project will exactly pay out over its economic life. In schedule F of the example, it can be seen from the cumulative cash flow that the project will pay out during the fifth year—by interpolation, in about 4.7 years. The reciprocal of this payout time gives another (though poor) measure of investment desirability—about 21 percent in this case. The profitability model can be used on a project-, product-, plant-, or company-wide basis. Its usefulness can sometimes be enhanced by assigning probabilities to some of the variables, and it can be used in developing optimum price-cost-volume relationships. Although the example described in schedules A–F pertains to a manufactured product, the same general format is adequate for most other types of projects encountered by the engineer.

Many management people are reluctant to acknowledge that such an analytical approach is worthwhile in business economics—perhaps questioning the accuracy implied by any mathematically expressed economic relationships. The real question is not whether this method is infallible, but simply whether there is any other approach as good in formulating assumptions and in evaluating the sensitivity of profits to changes in the variables.

Perhaps the main virtue of this approach is that it requires a great deal of objective thought and study for its preparation. Making the engineer carefully think the project through and state all of his assumptions minimizes the danger of having an individual's preconceived notions lead, after much analysis, only to foregone conclusions.

SUMMARY

The economic or financial model brings together, in summary form, all of the many factors relating to the economic success of a business venture. Profit, a function of price, cost, and volume, is the primary objective of business, so the model necessarily encompasses all the elements that affect price, cost, and volume. The price-volume relationship characterizes the market for a given product and determines the number of units that can be sold at various price levels; the cost-volume relationship identifies economies of scale in production and establishes the costs for whatever level of production is decided upon; and the price-cost relationship defines the net profit margin per unit produced and sold. A comprehensive financial model for a company or for a new venture incorporates all these relationships in separate submodels relating to income, costs, capital requirements, depreciation, profits, and cash flows, which in turn are used to develop appropriate measures of profitability. By relating all the significant factors in a series of mathematical expressions, the impact of changes in any given factor on other factors, and on the overall project's economic outcomes, can be determined. The model permits individual, controllable factors to be varied so that optimum results can be obtained from the project as a whole.

APPENDIX

Interest Tables

The following interest tables cover 0–25 percent in 1.0 percent increments, for time periods from 1 to 40 years, for the following factors:

Single payment compound amount factor.

Single payment present worth factor.

Sinking fund factor for uniform annual series.

Capital recovery factor for uniform annual series.

Compound amount factor for uniform annual series.

Present worth factor for uniform annual series.

The following notation has been used: I represents the interest rate per period; N represents the number of interest periods; P represents a present sum of money; S represents a sum of money N interest periods from the present time that, when discounted at interest rate I, is equivalent to a present amount P; R represents the uniform end-of-period payment for N time periods that, discounted at interest rate I, makes the series equivalent to a present amount P.

These tables are reproduced by permission from *Interest Tables: 0 to 25 Percent*, published by the Competitive Service Committee of the Edison Electric Institute (EEI Publ. No. 67-21).

0.00 PERCENT COMPOUND INTEREST FACTORS

N PERIODS	------SINGLE PAYMENT------		------UNIFORM ANNUAL SERIES------			
	COMPOUND AMOUNT FACTOR GIVEN P TO FIND S	PRESENT WORTH FACTOR GIVEN S TO FIND P	SINKING FUND FACTOR GIVEN S TO FIND R	CAPITAL RECOVERY FACTOR GIVEN P TO FIND R	COMPOUND AMOUNT FACTOR GIVEN R TO FIND S	PRESENT WORTH FACTOR GIVEN R TO FIND P
	$(1 + I)^{**N}$	$\dfrac{1}{(1 + I)^{**N}}$	$\dfrac{I}{(1 + I)^{**N} - 1}$	$\dfrac{I(1 + I)^{**N}}{(1 + I)^{**N} - 1}$	$\dfrac{(1 + I)^{**N} - 1}{I}$	$\dfrac{(1 + I)^{**N} - 1}{I(1 + I)^{**N}}$

Equals 1.0 for "N" Periods.

Equals 1.0 for "N" Periods.

Equals 1.0 divided by "N" Periods.

Equals 1.0 divided by "N" Periods.

Equals 1.0 times "N" Periods.

Equals 1.0 times "N" Periods.

1.00 PERCENT COMPOUND INTEREST FACTORS

	SINGLE PAYMENT			UNIFORM ANNUAL SERIES			
N PERIODS	COMPOUND AMOUNT FACTOR GIVEN P TO FIND S	PRESENT WORTH FACTOR GIVEN S TO FIND P	SINKING FUND FACTOR GIVEN S TO FIND R	CAPITAL RECOVERY FACTOR GIVEN P TO FIND R	COMPOUND AMOUNT FACTOR GIVEN R TO FIND S	PRESENT WORTH FACTOR GIVEN R TO FIND P	N PERIODS
	$(1 + I)^{**N}$	$\dfrac{1}{(1 + I)^{**N}}$	$\dfrac{I}{(1 + I)^{**N} - 1}$	$\dfrac{I(1 + I)^{**N}}{(1 + I)^{**N} - 1}$	$\dfrac{(1 + I)^{**N} - 1}{I}$	$\dfrac{(1 + I)^{**N} - 1}{I(1 + I)^{**N}}$	
1	1.0100000	.9900990	1.0000000	1.0100000	1.0000000	.9900990	1
2	1.0201000	.9802960	.4975124	.5075124	2.0100000	1.9703951	2
3	1.0303010	.9705901	.3300221	.3400221	3.0301000	2.9409852	3
4	1.0406040	.9609803	.2462811	.2562811	4.0604010	3.9019656	4
5	1.0510101	.9514657	.1960398	.2060398	5.1010050	4.8534312	5
6	1.0615202	.9420452	.1625484	.1725484	6.1520151	5.7954765	6
7	1.0721354	.9327181	.1386283	.1486283	7.2135352	6.7281945	7
8	1.0828567	.9234832	.1206903	.1306903	8.2856706	7.6516778	8
9	1.0936853	.9143398	.1067404	.1167404	9.3685273	8.5660176	9
10	1.1046221	.9052870	.0955821	.1055821	10.4622125	9.4713045	10
11	1.1156683	.8963237	.0864541	.0964541	11.5668347	10.3676282	11
12	1.1268250	.8874492	.0788488	.0888488	12.6825030	11.2550775	12
13	1.1380933	.8786626	.0724148	.0824148	13.8093280	12.1337401	13
14	1.1494742	.8699630	.0669012	.0769012	14.9474213	13.0037030	14
15	1.1609690	.8613495	.0621238	.0721238	16.0968955	13.8650525	15

N							N
16	1.1725786	.8528213	.0579446	.0679446	17.2578645	14.7178738	16
17	1.1843044	.8443775	.0542581	.0642581	18.4304431	15.5622513	17
18	1.1961475	.8360173	.0509820	.0609820	19.6147476	16.3982686	18
19	1.2081090	.8277399	.0480518	.0580518	20.8108950	17.2260085	19
20	1.2201900	.8195445	.0454153	.0554153	22.0190040	18.0455530	20
21	1.2323919	.8114302	.0430308	.0530308	23.2391940	18.8569831	21
22	1.2447159	.8033962	.0408637	.0508637	24.4715860	19.6603793	22
23	1.2571630	.7954418	.0388858	.0488858	25.7163018	20.4558211	23
24	1.2697346	.7875661	.0370735	.0470735	26.9734649	21.2433873	24
25	1.2824320	.7797684	.0354068	.0454068	28.2431995	22.0231557	25
26	1.2952563	.7720480	.0338689	.0438689	29.5256315	22.7952037	26
27	1.3082089	.7644039	.0324455	.0424455	30.8206878	23.5596076	27
28	1.3212910	.7568356	.0311244	.0411244	32.1290967	24.3164432	28
29	1.3345039	.7493421	.0298950	.0398950	33.4503877	25.0657853	29
30	1.3478489	.7419229	.0287481	.0387481	34.7848915	25.8077082	30
31	1.3613274	.7345771	.0276757	.0376757	36.1327404	26.5422854	31
32	1.3749407	.7273041	.0266709	.0366709	37.4940679	27.2695895	32
33	1.3886901	.7201031	.0257274	.0357274	38.8690085	27.9896925	33
34	1.4025770	.7129733	.0248400	.0348400	40.2576986	28.7026659	34
35	1.4166028	.7059142	.0240037	.0340037	41.6602756	29.4085801	35
36	1.4307688	.6989249	.0232143	.0332143	43.0768784	30.1075050	36
37	1.4450765	.6920049	.0224680	.0324680	44.5076471	30.7995099	37
38	1.4595272	.6851534	.0217615	.0317615	45.9527236	31.4846633	38
39	1.4741225	.6783697	.0210916	.0310916	47.4122509	32.1630330	39
40	1.4888637	.6716531	.0204556	.0304556	48.8863734	32.8346861	40

NOTE- **N IS EXPONENT N

273

2.00 PERCENT COMPOUND INTEREST FACTORS

	-----SINGLE PAYMENT-----			-----UNIFORM ANNUAL SERIES-----			
N PERIODS	COMPOUND AMOUNT FACTOR GIVEN P TO FIND S $(1 + I)**N$	PRESENT WORTH FACTOR GIVEN S TO FIND P $\dfrac{1}{(1 + I)**N}$	SINKING FUND FACTOR GIVEN S TO FIND R $\dfrac{I}{(1 + I)**N - 1}$	CAPITAL RECOVERY FACTOR GIVEN P TO FIND R $\dfrac{I(1 + I)**N}{(1 + I)**N - 1}$	COMPOUND AMOUNT FACTOR GIVEN R TO FIND S $\dfrac{(1 + I)**N - 1}{I}$	PRESENT WORTH FACTOR GIVEN R TO FIND P $\dfrac{(1 + I)**N - 1}{I(1 + I)**N}$	N PERIODS
1	1.0200000	.9803922	1.0000000	1.0200000	1.0000000	.9803922	1
2	1.0404000	.9611688	.4950495	.5150495	2.0200000	1.9415609	2
3	1.0612080	.9423223	.3267547	.3467547	3.0604000	2.8838833	3
4	1.0824322	.9238454	.2426238	.2626238	4.1216080	3.8077287	4
5	1.1040808	.9057308	.1921584	.2121584	5.2040402	4.7134595	5
6	1.1261624	.8879714	.1585258	.1785258	6.3081210	5.6014309	6
7	1.1486857	.8705602	.1345120	.1545120	7.4342834	6.4719911	7
8	1.1716594	.8534904	.1165098	.1365098	8.5829691	7.3254814	8
9	1.1950926	.8367553	.1025154	.1225154	9.7546284	8.1622367	9
10	1.2189944	.8203483	.0913265	.1113265	10.9497210	8.9825850	10
11	1.2433743	.8042630	.0821779	.1021779	12.1687154	9.7868480	11
12	1.2682418	.7884932	.0745596	.0945596	13.4120897	10.5753412	12
13	1.2936066	.7730325	.0681184	.0881184	14.6803315	11.3483737	13
14	1.3194788	.7578750	.0626020	.0826020	15.9739382	12.1062488	14
15	1.3458683	.7430147	.0578255	.0778255	17.2934169	12.8492635	15

n						n
16	1.3727857	.7284458	.0536501	.0736501	18.6392853	13.5777093	16
17	1.4002414	.7141626	.0499698	.0699698	20.0120710	14.2916719	17
18	1.4282462	.7001594	.0467021	.0667021	21.4123124	14.9920313	18
19	1.4568112	.6864308	.0437818	.0637818	22.8405586	15.6784620	19
20	1.4859474	.6729713	.0411567	.0611567	24.2973698	16.3514333	20
21	1.5156663	.6597758	.0387848	.0587848	25.7833172	17.0122092	21
22	1.5459797	.6468390	.0366314	.0566314	27.2989835	17.6580482	22
23	1.5768993	.6341559	.0346681	.0546681	28.8449632	18.2922041	23
24	1.6084372	.6217215	.0328711	.0528711	30.4218625	18.9139256	24
25	1.6406060	.6095309	.0312204	.0512204	32.0302997	19.5234565	25
26	1.6734181	.5975793	.0296992	.0496992	33.6709057	20.1210358	26
27	1.7068865	.5858620	.0282931	.0482931	35.3443238	20.7068978	27
28	1.7410242	.5743746	.0269897	.0469897	37.0512103	21.2812724	28
29	1.7758447	.5631123	.0257784	.0457784	38.7923345	21.8443847	29
30	1.8113616	.5520709	.0246499	.0446499	40.5680792	22.3964556	30
31	1.8475888	.5412460	.0235963	.0435963	42.3794408	22.9377015	31
32	1.8845406	.5306333	.0226106	.0426106	44.2270296	23.4683348	32
33	1.9222314	.5202287	.0216865	.0416865	46.1115702	23.9885636	33
34	1.9606760	.5100282	.0208187	.0408187	48.0338016	24.4985917	34
35	1.9998896	.5000276	.0200022	.0400022	49.9944776	24.9986193	35
36	2.0398873	.4902232	.0192329	.0392329	51.9943672	25.4888425	36
37	2.0806851	.4806109	.0185068	.0385068	54.0342545	25.9694534	37
38	2.1222988	.4711872	.0178206	.0378206	56.1149396	26.4406406	38
39	2.1647448	.4619482	.0171711	.0371711	58.2372384	26.9025888	39
40	2.2080397	.4528904	.0165557	.0365557	60.4019832	27.3554792	40

NOTE— **N IS EXPONENT N

3.00 PERCENT COMPOUND INTEREST FACTORS

N PERIODS	SINGLE PAYMENT		UNIFORM ANNUAL SERIES				N PERIODS
	COMPOUND AMOUNT FACTOR GIVEN P TO FIND S	PRESENT WORTH FACTOR GIVEN S TO FIND P	SINKING FUND FACTOR GIVEN S TO FIND R	CAPITAL RECOVERY FACTOR GIVEN P TO FIND R	COMPOUND AMOUNT FACTOR GIVEN R TO FIND S	PRESENT WORTH FACTOR GIVEN R TO FIND P	
	$(1 + I)**N$	$\dfrac{1}{(1 + I)**N}$	$\dfrac{I}{(1 + I)**N - 1}$	$\dfrac{I(1 + I)**N}{(1 + I)**N - 1}$	$\dfrac{(1 + I)**N - 1}{I}$	$\dfrac{(1 + I)**N - 1}{I(1 + I)**N}$	
1	1.0300000	.9708738	1.0000000	1.0300000	1.0000000	.9708738	1
2	1.0609000	.9425959	.4926108	.5226108	2.0300000	1.9134697	2
3	1.0927270	.9151417	.3235304	.3535304	3.0909000	2.8286114	3
4	1.1255088	.8884870	.2390270	.2690270	4.1836270	3.7170984	4
5	1.1592741	.8626088	.1883546	.2183546	5.3091358	4.5797072	5
6	1.1940523	.8374843	.1545975	.1845975	6.4684099	5.4171914	6
7	1.2298739	.8130915	.1305064	.1605064	7.6624622	6.2302830	7
8	1.2667701	.7894092	.1124564	.1424564	8.8923360	7.0196922	8
9	1.3047732	.7664167	.0984339	.1284339	10.1591061	7.7861089	9
10	1.3439164	.7440939	.0872305	.1172305	11.4638793	8.5302028	10
11	1.3842339	.7224213	.0780774	.1080774	12.8077957	9.2526241	11
12	1.4257609	.7013799	.0704621	.1004621	14.1920296	9.9540040	12
13	1.4685337	.6809513	.0640295	.0940295	15.6177904	10.6349553	13
14	1.5125897	.6611178	.0585263	.0885263	17.0863242	11.2960731	14
15	1.5579674	.6418619	.0537666	.0837666	18.5989139	11.9379351	15

n							n
16	1.6047064	.6231669	.0496108	.0796108	20.1568813	12.5611020	16
17	1.6528476	.6050164	.0459525	.0759525	21.7615877	13.1661185	17
18	1.7024331	.5873946	.0427087	.0727087	23.4144354	13.7535131	18
19	1.7535061	.5702860	.0398139	.0698139	25.1168684	14.3237991	19
20	1.8061112	.5536758	.0372157	.0672157	26.8703745	14.8774749	20
21	1.8602946	.5375493	.0348718	.0648718	28.6764857	15.4150241	21
22	1.9161034	.5218925	.0327474	.0627474	30.5367803	15.9369166	22
23	1.9735865	.5066917	.0308139	.0608139	32.4528837	16.4436084	23
24	2.0327941	.4919337	.0290474	.0590474	34.4264702	16.9355421	24
25	2.0937779	.4776056	.0274279	.0574279	36.4592643	17.4131477	25
26	2.1565913	.4636947	.0259383	.0559383	38.5530423	17.8768424	26
27	2.2212890	.4501891	.0245642	.0545642	40.7096335	18.3270315	27
28	2.2879277	.4370768	.0232932	.0532932	42.9309225	18.7641082	28
29	2.3565655	.4243464	.0221147	.0521147	45.2188502	19.1884546	29
30	2.4272625	.4119868	.0210193	.0510193	47.5754157	19.6004413	30
31	2.5000803	.3999871	.0199989	.0499989	50.0026782	20.0004285	31
32	2.5750828	.3883370	.0190466	.0490466	52.5027585	20.3887655	32
33	2.6523352	.3770262	.0181561	.0481561	55.0778413	20.7657918	33
34	2.7319053	.3660449	.0173220	.0473220	57.7301765	21.1318367	34
35	2.8138625	.3553834	.0165393	.0465393	60.4620818	21.4872201	35
36	2.8982783	.3450324	.0158038	.0458036	63.2759443	21.8322525	36
37	2.9852267	.3349829	.0151116	.0451116	66.1742226	22.1672354	37
38	3.0747835	.3252262	.0144593	.0444593	69.1594493	22.4924616	38
39	3.1670270	.3157535	.0138439	.0438439	72.2342328	22.8082151	39
40	3.2620378	.3065568	.0132624	.0432624	75.4012597	23.1147720	40

NOTE- **N IS EXPONENT N

4.00 PERCENT COMPOUND INTEREST FACTORS

	----SINGLE PAYMENT----		----UNIFORM ANNUAL SERIES----				
N PERIODS	COMPOUND AMOUNT FACTOR GIVEN P TO FIND S $(1 + I)**N$	PRESENT WORTH FACTOR GIVEN S TO FIND P $\dfrac{1}{(1 + I)**N}$	SINKING FUND FACTOR GIVEN S TO FIND R $\dfrac{I}{(1 + I)**N - 1}$	CAPITAL RECOVERY FACTOR GIVEN P TO FIND R $\dfrac{I(1 + I)**N}{(1 + I)**N - 1}$	COMPOUND AMOUNT FACTOR GIVEN R TO FIND S $\dfrac{(1 + I)**N - 1}{I}$	PRESENT WORTH FACTOR GIVEN R TO FIND P $\dfrac{(1 + I)**N - 1}{I(1 + I)**N}$	N PERIODS
1	1.0400000	.9615385	1.0000000	1.0400000	1.0000000	.9615385	1
2	1.0816000	.9245562	.4901961	.5301961	2.0400000	1.8860947	2
3	1.1248640	.8889964	.3203485	.3603485	3.1216000	2.7750910	3
4	1.1698586	.8548042	.2354900	.2754900	4.2464640	3.6298952	4
5	1.2166529	.8219271	.1846271	.2246271	5.4163226	4.4518223	5
6	1.2653190	.7903145	.1507619	.1907619	6.6329755	5.2421369	6
7	1.3159318	.7599178	.1266096	.1666096	7.8982945	6.0020547	7
8	1.3685691	.7306902	.1085278	.1485278	9.2142263	6.7327449	8
9	1.4233118	.7025867	.0944930	.1344930	10.5827953	7.4353316	9
10	1.4802443	.6755642	.0832909	.1232909	12.0061071	8.1108958	10
11	1.5394541	.6495809	.0741490	.1141490	13.4863514	8.7604767	11
12	1.6010322	.6245970	.0665522	.1065522	15.0258055	9.3850738	12
13	1.6650735	.6005741	.0601437	.1001437	16.6268377	9.9856478	13
14	1.7316764	.5774751	.0546690	.0946690	18.2919112	10.5631229	14
15	1.8009435	.5552665	.0499411	.0899411	20.0235876	11.1183874	15

n							n
16	1.8729812	.5339082	.0458200	21.8245311	.0858200	11.6522956	16
17	1.9479005	.5133732	.0421985	23.6975124	.0821985	12.1656689	17
18	2.0258165	.4936281	.0389933	25.6454129	.0789933	12.6592970	18
19	2.1068492	.4746424	.0361386	27.6712294	.0761386	13.1339394	19
20	2.1911231	.4563869	.0335818	29.7780786	.0735818	13.5903263	20
21	2.2787681	.4388336	.0312801	31.9692017	.0712801	14.0291599	21
22	2.3699188	.4219554	.0291988	34.2479698	.0691988	14.4511153	22
23	2.4647155	.4057263	.0273091	36.6178886	.0673091	14.8568417	23
24	2.5633042	.3901215	.0255868	39.0826041	.0655868	15.2469631	24
25	2.6658363	.3751168	.0240120	41.6459083	.0640120	15.6220799	25
26	2.7724698	.3606892	.0225674	44.3117446	.0625674	15.9827692	26
27	2.8833686	.3468166	.0212385	47.0842144	.0612385	16.3295857	27
28	2.9987033	.3334775	.0200130	49.9675830	.0600130	16.6630632	28
29	3.1186515	.3206514	.0188799	52.9662863	.0588799	16.9837146	29
30	3.2433975	.3083187	.0178301	56.0849378	.0578301	17.2920333	30
31	3.3731334	.2964603	.0168554	59.3283353	.0568554	17.5884936	31
32	3.5080587	.2850579	.0159486	62.7014687	.0559486	17.8735515	32
33	3.6483811	.2740942	.0151036	66.2095274	.0551036	18.1476457	33
34	3.7943163	.2635521	.0143148	69.8579085	.0543148	18.4111978	34
35	3.9460890	.2534155	.0135773	73.6522249	.0535773	18.6646132	35
36	4.1039326	.2436687	.0128869	77.5981138	.0528869	18.9082820	36
37	4.2680899	.2342968	.0122396	81.7022464	.0522396	19.1425788	37
38	4.4388135	.2252854	.0116319	85.9703363	.0516319	19.3678642	38
39	4.6163660	.2166206	.0110608	90.4091497	.0510608	19.5844848	39
40	4.8010206	.2082890	.0105235	95.0255157	.0505235	19.7927739	40

NOTE— **N IS EXPONENT N

5.00 PERCENT COMPOUND INTEREST FACTORS

	SINGLE PAYMENT			UNIFORM ANNUAL SERIES			
N PERIODS	COMPOUND AMOUNT FACTOR GIVEN P TO FIND S $(1 + I)^{**N}$	PRESENT WORTH FACTOR GIVEN S TO FIND P $\dfrac{1}{(1 + I)^{**N}}$	SINKING FUND FACTOR GIVEN S TO FIND R $\dfrac{I}{(1 + I)^{**N} - 1}$	CAPITAL RECOVERY FACTOR GIVEN P TO FIND R $\dfrac{I(1 + I)^{**N}}{(1 + I)^{**N} - 1}$	COMPOUND AMOUNT FACTOR GIVEN R TO FIND S $\dfrac{(1 + I)^{**N} - 1}{I}$	PRESENT WORTH FACTOR GIVEN R TO FIND P $\dfrac{(1 + I)^{**N} - 1}{I(1 + I)^{**N}}$	N PERIODS
1	1.0500000	.9523810	1.0000000	1.0500000	1.0000000	.9523810	1
2	1.1025000	.9070295	.4878049	.5378049	2.0500000	1.8594104	2
3	1.1576250	.8638376	.3172086	.3672086	3.1525000	2.7232480	3
4	1.2155063	.8227025	.2320118	.2820118	4.3101250	3.5459505	4
5	1.2762816	.7835262	.1809748	.2309748	5.5256313	4.3294767	5
6	1.3400956	.7462154	.1470175	.1970175	6.8019128	5.0756921	6
7	1.4071004	.7106813	.1228198	.1728198	8.1420085	5.7863734	7
8	1.4774554	.6768394	.1047218	.1547218	9.5491089	6.4632128	8
9	1.5513282	.6446089	.0906901	.1406901	11.0265643	7.1078217	9
10	1.6288946	.6139133	.0795046	.1295046	12.5778925	7.7217349	10
11	1.7103394	.5846793	.0703889	.1203889	14.2067872	8.3064142	11
12	1.7958563	.5568374	.0628254	.1128254	15.9171265	8.8632516	12
13	1.8856491	.5303214	.0564558	.1064558	17.7129828	9.3935730	13
14	1.9799316	.5050680	.0510240	.1010240	19.5986320	9.8986409	14
15	2.0789282	.4810171	.0463423	.0963423	21.5785636	10.3796580	15

n						n
16	2.1828746	.4581115	.0422699	.0922699	23.6574918	10.8377696	16
17	2.2920183	.4362967	.0386991	.0886991	25.8403664	11.2740662	17
18	2.4066192	.4155207	.0355462	.0855462	28.1323847	11.6895869	18
19	2.5269502	.3957340	.0327450	.0827450	30.5390039	12.0853209	19
20	2.6532977	.3768895	.0302426	.0802426	33.0659541	12.4622103	20
21	2.7859626	.3589424	.0279961	.0779961	35.7192518	12.8211527	21
22	2.9252607	.3418499	.0259705	.0759705	38.5052144	13.1630026	22
23	3.0715238	.3255713	.0241368	.0741368	41.4304751	13.4885739	23
24	3.2250999	.3100679	.0224709	.0724709	44.5019989	13.7986418	24
25	3.3863549	.2953028	.0209525	.0709525	47.7270988	14.0939446	25
26	3.5556727	.2812407	.0195643	.0695643	51.1134538	14.3751853	26
27	3.7334563	.2678483	.0182919	.0682919	54.6691264	14.6430336	27
28	3.9201291	.2550936	.0171225	.0671225	58.4025828	14.8981273	28
29	4.1161356	.2429463	.0160455	.0660455	62.3227119	15.1410736	29
30	4.3219424	.2313774	.0150514	.0650514	66.4388475	15.3724510	30
31	4.5380395	.2203595	.0141321	.0641321	70.7607899	15.5928105	31
32	4.7649415	.2098662	.0132804	.0632804	75.2988294	15.8026767	32
33	5.0031885	.1998725	.0124900	.0624900	80.0637708	16.0025492	33
34	5.2533480	.1903548	.0117554	.0617554	85.0669594	16.1929040	34
35	5.5160154	.1812903	.0110717	.0610717	90.3203074	16.3741943	35
36	5.7918161	.1726574	.0104345	.0604345	95.8363227	16.5468517	36
37	6.0814069	.1644356	.0098398	.0598398	101.6281389	16.7112873	37
38	6.3854773	.1566054	.0092842	.0592842	107.7095458	16.8678927	38
39	6.7047512	.1491480	.0087646	.0587646	114.0950231	17.0170407	39
40	7.0399887	.1420457	.0082782	.0582782	120.7997742	17.1590864	40

NOTE- **N IS EXPONENT N

6.00 PERCENT COMPOUND INTEREST FACTORS

N PERIODS	SINGLE PAYMENT COMPOUND AMOUNT FACTOR GIVEN P TO FIND S $(1+I)^{**N}$	PRESENT WORTH FACTOR GIVEN S TO FIND P $\frac{1}{(1+I)^{**N}}$	SINKING FUND FACTOR GIVEN S TO FIND R $\frac{I}{(1+I)^{**N}-1}$	UNIFORM ANNUAL SERIES CAPITAL RECOVERY FACTOR GIVEN P TO FIND R $\frac{I(1+I)^{**N}}{(1+I)^{**N}-1}$	COMPOUND AMOUNT FACTOR GIVEN R TO FIND S $\frac{(1+I)^{**N}-1}{I}$	PRESENT WORTH FACTOR GIVEN R TO FIND P $\frac{(1+I)^{**N}-1}{I(1+I)^{**N}}$	N PERIODS
1	1.0600000	.9433962	1.0000000	1.0600000	1.0000000	.9433962	1
2	1.1236000	.8899964	.4854369	.5454369	2.0600000	1.8333927	2
3	1.1910160	.8396193	.3141098	.3741098	3.1836000	2.6730119	3
4	1.2624770	.7920937	.2285915	.2885915	4.3746160	3.4651056	4
5	1.3382256	.7472582	.1773964	.2373964	5.6370930	4.2123638	5
6	1.4185191	.7049605	.1433626	.2033626	6.9753185	4.9173243	6
7	1.5036303	.6650571	.1191350	.1791350	8.3938376	5.5823814	7
8	1.5938481	.6274124	.1010359	.1610359	9.8974679	6.2097938	8
9	1.6894790	.5918985	.0870222	.1470222	11.4913160	6.8016923	9
10	1.7908477	.5583948	.0758680	.1358680	13.1807949	7.3600871	10
11	1.8982986	.5267875	.0667929	.1267929	14.9716426	7.8868746	11
12	2.0121965	.4969694	.0592770	.1192770	16.8699412	8.3838439	12
13	2.1329283	.4688390	.0529601	.1129601	18.8821377	8.8526830	13
14	2.2609040	.4423010	.0475849	.1075849	21.0150659	9.2949839	14
15	2.3965582	.4172651	.0429628	.1029628	23.2759699	9.7122490	15

n							n
16	10.1058953	25.6725281	.0989521	.0389521	.3936463	2.5403517	16
17	10.4772597	28.2128798	.0954448	.0354448	.3713644	2.6927728	17
18	10.8276035	30.9056525	.0923565	.0323565	.3503438	2.8543392	18
19	11.1581165	33.7599917	.0896209	.0296209	.3305130	3.0255995	19
20	11.4699212	36.7855912	.0871846	.0271846	.3118047	3.2071355	20
21	11.7640766	39.9927267	.0850045	.0250045	.2941554	3.3995636	21
22	12.0415817	43.3922903	.0830456	.0230456	.2775051	3.6035374	22
23	12.3033790	46.9958277	.0812785	.0212785	.2617973	3.8197497	23
24	12.5503575	50.8155774	.0796790	.0196790	.2469785	4.0489346	24
25	12.7833562	54.8645120	.0782267	.0182267	.2329986	4.2918707	25
26	13.0031662	59.1563827	.0769043	.0169043	.2198100	4.5493830	26
27	13.2105341	63.7057657	.0756972	.0156972	.2073680	4.8223459	27
28	13.4061643	68.5281116	.0745926	.0145926	.1956301	5.1116867	28
29	13.5907210	73.6397983	.0735796	.0135796	.1845567	5.4183879	29
30	13.7648312	79.0581862	.0726489	.0126489	.1741101	5.7434912	30
31	13.9290860	84.8016774	.0717922	.0117922	.1642548	6.0881006	31
32	14.0840434	90.8897780	.0710023	.0110023	.1549574	6.4533867	32
33	14.2302296	97.3431647	.0702729	.0102729	.1461862	6.8405899	33
34	14.3681411	104.1837546	.0695984	.0095984	.1379115	7.2510253	34
35	14.4982464	111.4347799	.0689739	.0089739	.1301052	7.6860868	35
36	14.6209871	119.1208667	.0683948	.0083948	.1227408	8.1472520	36
37	14.7367803	127.2681187	.0678574	.0078574	.1157932	8.6360871	37
38	14.8460192	135.9042058	.0673581	.0073581	.1092389	9.1542523	38
39	14.9490747	145.0584581	.0668938	.0068938	.1030555	9.7035075	39
40	15.0462969	154.7619656	.0664615	.0064615	.0972222	10.2857179	40

NOTE- **N IS EXPONENT N

7.00 PERCENT COMPOUND INTEREST FACTORS

N PERIODS	SINGLE PAYMENT COMPOUND AMOUNT FACTOR GIVEN P TO FIND S $(1 + I)**N$	SINGLE PAYMENT PRESENT WORTH FACTOR GIVEN S TO FIND P $\frac{1}{(1 + I)**N}$	UNIFORM ANNUAL SERIES SINKING FUND FACTOR GIVEN S TO FIND R $\frac{I}{(1 + I)**N - 1}$	UNIFORM ANNUAL SERIES CAPITAL RECOVERY FACTOR GIVEN P TO FIND R $\frac{I(1 + I)**N}{(1 + I)**N - 1}$	UNIFORM ANNUAL SERIES COMPOUND AMOUNT FACTOR GIVEN R TO FIND S $\frac{(1 + I)**N - 1}{I}$	UNIFORM ANNUAL SERIES PRESENT WORTH FACTOR GIVEN R TO FIND P $\frac{(1 + I)**N - 1}{I(1 + I)**N}$	N PERIODS
1	1.0700000	.9345794	1.0000000	1.0700000	1.0000000	.9345794	1
2	1.1449000	.8734387	.4830918	.5530918	2.0700000	1.8080182	2
3	1.2250430	.8162979	.3110517	.3810517	3.2149000	2.6243160	3
4	1.3107960	.7628952	.2252281	.2952281	4.4399430	3.3872113	4
5	1.4025517	.7129862	.1738907	.2438907	5.7507390	4.1001974	5
6	1.5007304	.6663422	.1397958	.2097958	7.1532907	4.7665397	6
7	1.6057815	.6227497	.1155532	.1855532	8.6540211	5.3892894	7
8	1.7181862	.5820091	.0974678	.1674678	10.2598026	5.9712985	8
9	1.8384592	.5439337	.0834865	.1534865	11.9779887	6.5152322	9
10	1.9671514	.5083493	.0723775	.1423775	13.8164480	7.0235815	10
11	2.1048520	.4750928	.0633569	.1333569	15.7835993	7.4986743	11
12	2.2521916	.4440120	.0559020	.1259020	17.8884513	7.9426863	12
13	2.4098450	.4149644	.0496508	.1196508	20.1406429	8.3576507	13
14	2.5785342	.3878172	.0443449	.1143449	22.5504879	8.7454680	14
15	2.7590315	.3624460	.0397946	.1097946	25.1290220	9.1079140	15

284

n						
16	2.9521637	.3387346	.0358576	.1058576	27.8880536	9.4466486
17	3.1588152	.3165744	.0324252	.1024252	30.8402173	9.7632230
18	3.3799323	.2958639	.0294126	.0994126	33.9990325	10.0590869
19	3.6165275	.2765083	.0267530	.0967530	37.3789648	10.3355952
20	3.8696845	.2584190	.0243929	.0943929	40.9954923	10.5940142
21	4.1405624	.2415131	.0222890	.0922890	44.8651768	10.8355273
22	4.4304017	.2257132	.0204058	.0904058	49.0057392	11.0612405
23	4.7405299	.2109469	.0187139	.0887139	53.4361409	11.2721874
24	5.0723670	.1971466	.0171890	.0871890	58.1766708	11.4693340
25	5.4274326	.1842492	.0158105	.0858105	63.2490377	11.6535832
26	5.8073529	.1721955	.0145610	.0845610	68.6764704	11.8257787
27	6.2138676	.1609304	.0134257	.0834257	74.4838233	11.9867090
28	6.6488384	.1504022	.0123919	.0823919	80.6976909	12.1371113
29	7.1142570	.1405628	.0114487	.0814487	87.3465293	12.2776741
30	7.6122550	.1313671	.0105864	.0805864	94.4607863	12.4090412
31	8.1451129	.1227730	.0097969	.0797969	102.0730414	12.5318142
32	8.7152708	.1147411	.0090729	.0790729	110.2181543	12.6465553
33	9.3253398	.1072347	.0084081	.0784081	118.9334251	12.7537900
34	9.9781135	.1002193	.0077967	.0777967	128.2587648	12.8540094
35	10.6765815	.0936629	.0072340	.0772340	138.2368784	12.9476723
36	11.4239422	.0875355	.0067153	.0767153	148.9134598	13.0352078
37	12.2236181	.0818088	.0062368	.0762368	160.3374020	13.1170166
38	13.0797714	.0764569	.0057951	.0757951	172.5610202	13.1934735
39	13.9948204	.0714550	.0053868	.0753868	185.6402916	13.2649285
40	14.9744578	.0667804	.0050091	.0750091	199.6351120	13.3317088

NOTE- **N IS EXPONENT N

8.00 PERCENT COMPOUND INTEREST FACTORS

	------SINGLE PAYMENT------		------UNIFORM ANNUAL SERIES------				
N PERIODS	COMPOUND AMOUNT FACTOR GIVEN P TO FIND S	PRESENT WORTH FACTOR GIVEN S TO FIND P	SINKING FUND FACTOR GIVEN S TO FIND R	CAPITAL RECOVERY FACTOR GIVEN P TO FIND R	COMPOUND AMOUNT FACTOR GIVEN R TO FIND S	PRESENT WORTH FACTOR GIVEN R TO FIND P	N PERIODS
	$(1 + I)**N$	$\dfrac{1}{(1 + I)**N}$	$\dfrac{I}{(1 + I)**N - 1}$	$\dfrac{I(1 + I)**N}{(1 + I)**N - 1}$	$\dfrac{(1 + I)**N - 1}{I}$	$\dfrac{(1 + I)**N - 1}{I(1 + I)**N}$	
1	1.0800000	.9259259	1.0000000	1.0800000	1.0000000	.9259259	1
2	1.1664000	.8573388	.4807692	.5607692	2.0800000	1.7832647	2
3	1.2597120	.7938322	.3080335	.3880335	3.2464000	2.5770970	3
4	1.3604890	.7350299	.2219208	.3019208	4.5061120	3.3121268	4
5	1.4693281	.6805832	.1704565	.2504565	5.8666010	3.9927100	5
6	1.5868743	.6301696	.1363154	.2163154	7.3359290	4.6228797	6
7	1.7138243	.5834904	.1120724	.1920724	8.9228034	5.2063701	7
8	1.8509302	.5402689	.0940148	.1740148	10.6366276	5.7466389	8
9	1.9990046	.5002490	.0800797	.1600797	12.4875578	6.2468879	9
10	2.1589250	.4631935	.0690295	.1490295	14.4865625	6.7100814	10
11	2.3316390	.4288829	.0600763	.1400763	16.6454875	7.1389643	11
12	2.5181701	.3971138	.0526950	.1326950	18.9771265	7.5360780	12
13	2.7196237	.3676979	.0465218	.1265218	21.4952966	7.9037759	13
14	2.9371936	.3404610	.0412969	.1212969	24.2149203	8.2442370	14
15	3.1721691	.3152417	.0368295	.1168295	27.1521139	8.5594787	15

n							n
16	8.8513692	30.3242830	.1129769	.0329769	.2918905	3.4259426	16
17	9.1216381	33.7502257	.1096294	.0296294	.2702690	3.7000181	17
18	9.3718871	37.4502437	.1067021	.0267021	.2502490	3.9960195	18
19	9.6035992	41.4462632	.1041276	.0241276	.2317121	4.3157011	19
20	9.8181474	45.7619643	.1018522	.0218522	.2145482	4.6609571	20
21	10.0168032	50.4229214	.0998323	.0198323	.1986557	5.0338337	21
22	10.2007437	55.4567552	.0980321	.0180321	.1839405	5.4365404	22
23	10.3710589	60.8932956	.0964222	.0164222	.1703153	5.8714636	23
24	10.5287583	66.7647592	.0949780	.0149780	.1576993	6.3411807	24
25	10.6747762	73.1059400	.0936788	.0136788	.1460179	6.8484752	25
26	10.8099780	79.9544151	.0925071	.0125071	.1352018	7.3963532	26
27	10.9351648	87.3507684	.0914481	.0114481	.1251868	7.9880615	27
28	11.0510785	95.3388298	.0904889	.0104889	.1159137	8.6271064	28
29	11.1584060	103.9659362	.0896185	.0096185	.1073275	9.3172749	29
30	11.2577833	113.2832111	.0888274	.0088274	.0993773	10.0626569	30
31	11.3497994	123.3458680	.0881073	.0081073	.0920160	10.8676694	31
32	11.4349994	134.2135374	.0874508	.0074508	.0852000	11.7370830	32
33	11.5138884	145.9506204	.0868516	.0068516	.0788889	12.6760496	33
34	11.5869337	158.6266701	.0863041	.0063041	.0730453	13.6901336	34
35	11.6545682	172.3168037	.0858033	.0058033	.0676345	14.7853443	35
36	11.7171928	187.1021480	.0853447	.0053447	.0626246	15.9681718	36
37	11.7751785	203.0703198	.0849244	.0049244	.0579857	17.2456256	37
38	11.8288690	220.3159454	.0845389	.0045389	.0536905	18.6252756	38
39	11.8785824	238.9412210	.0841851	.0041851	.0497134	20.1152977	39
40	11.9246133	259.0565187	.0838602	.0038602	.0460309	21.7245215	40

NOTE— **N IS EXPONENT N

9.00 PERCENT COMPOUND INTEREST FACTORS

| | SINGLE PAYMENT | | UNIFORM ANNUAL SERIES | | | | |
N PERIODS	COMPOUND AMOUNT FACTOR GIVEN P TO FIND S $(1 + I)^{**N}$	PRESENT WORTH FACTOR GIVEN S TO FIND P $\dfrac{1}{(1 + I)^{**N}}$	SINKING FUND FACTOR GIVEN S TO FIND R $\dfrac{I}{(1 + I)^{**N} - 1}$	CAPITAL RECOVERY FACTOR GIVEN P TO FIND R $\dfrac{I(1 + I)^{**N}}{(1 + I)^{**N} - 1}$	COMPOUND AMOUNT FACTOR GIVEN R TO FIND S $\dfrac{(1 + I)^{**N} - 1}{I}$	PRESENT WORTH FACTOR GIVEN R TO FIND P $\dfrac{(1 + I)^{**N} - 1}{I(1 + I)^{**N}}$	N PERIODS
1	1.0900000	.9174312	1.0000000	1.0900000	1.0000000	.9174312	1
2	1.1881000	.8416800	.4784689	.5684689	2.0900000	1.7591112	2
3	1.2950290	.7721835	.3050548	.3950548	3.2781000	2.5312947	3
4	1.4115816	.7084252	.2186687	.3086687	4.5731290	3.2397199	4
5	1.5386240	.6499314	.1670925	.2570925	5.9847106	3.8896513	5
6	1.6771001	.5962673	.1329198	.2229198	7.5233346	4.4859186	6
7	1.8280391	.5470342	.1086905	.1986905	9.2004347	5.0329528	7
8	1.9925626	.5018663	.0906744	.1806744	11.0284738	5.5348191	8
9	2.1718933	.4604278	.0767988	.1667988	13.0210364	5.9952469	9
10	2.3673637	.4224108	.0658201	.1558201	15.1929297	6.4176577	10
11	2.5804264	.3875329	.0569467	.1469467	17.5602934	6.8051906	11
12	2.8126648	.3555347	.0496507	.1396507	20.1407198	7.1607253	12
13	3.0658046	.3261786	.0435666	.1335666	22.9553846	7.4869039	13
14	3.3417270	.2992465	.0384332	.1284332	26.0191892	7.7861504	14
15	3.6424825	.2745380	.0340589	.1240589	29.3609162	8.0606884	15

n							n
16	3.9703059	.2518698	.0302999	.1202999	33.0033987	8.3125582	16
17	4.3276334	.2310732	.0270462	.1170462	36.9737046	8.5436314	17
18	4.7171204	.2119937	.0242123	.1142123	41.3013380	8.7556251	18
19	5.1416613	.1944897	.0217304	.1117304	46.0184584	8.9501148	19
20	5.6044108	.1784309	.0195465	.1095465	51.1601196	9.1285457	20
21	6.1088077	.1636981	.0176166	.1076166	56.7645304	9.2922437	21
22	6.6586004	.1501817	.0159050	.1059050	62.8733381	9.4424254	22
23	7.2578745	.1377814	.0143819	.1043819	69.5319386	9.5802068	23
24	7.9110832	.1264049	.0130226	.1030226	76.7898131	9.7066118	24
25	8.6230807	.1159678	.0118063	.1018063	84.7008962	9.8225796	25
26	9.3391579	.1063925	.0107154	.1007154	93.3239769	9.9289721	26
27	10.2450821	.0976078	.0097349	.0997349	102.7231348	10.0265799	27
28	11.1671395	.0895484	.0088520	.0988520	112.9682169	10.1161284	28
29	12.1721821	.0821545	.0080557	.0980557	124.1353565	10.1982829	29
30	13.2676785	.0753711	.0073364	.0973364	136.3075385	10.2736540	30
31	14.4617695	.0691478	.0066856	.0966856	149.5752170	10.3428019	31
32	15.7633288	.0634384	.0060962	.0960962	164.0369865	10.4062403	32
33	17.1820284	.0582003	.0055617	.0955617	179.8003153	10.4644406	33
34	18.7284109	.0533948	.0050766	.0950766	196.9823437	10.5178354	34
35	20.4139679	.0489861	.0046358	.0946358	215.7107547	10.5668215	35
36	22.2512250	.0449413	.0042350	.0942350	236.1247226	10.6117628	36
37	24.2538353	.0412306	.0038703	.0938703	258.3759476	10.6529934	37
38	26.4366805	.0378262	.0035382	.0935382	282.6297829	10.6908196	38
39	28.8159817	.0347030	.0032356	.0932356	309.0664633	10.7255226	39
40	31.4094201	.0318376	.0029596	.0929596	337.8824450	10.7573602	40

NOTE- **N IS EXPONENT N

10.00 PERCENT COMPOUND INTEREST FACTORS

	SINGLE PAYMENT		UNIFORM ANNUAL SERIES				
N PERIODS	COMPOUND AMOUNT FACTOR GIVEN P TO FIND S $(1+I)^{**N}$	PRESENT WORTH FACTOR GIVEN S TO FIND P $\dfrac{1}{(1+I)^{**N}}$	SINKING FUND FACTOR GIVEN S TO FIND R $\dfrac{I}{(1+I)^{**N}-1}$	CAPITAL RECOVERY FACTOR GIVEN P TO FIND R $\dfrac{I(1+I)^{**N}}{(1+I)^{**N}-1}$	COMPOUND AMOUNT FACTOR GIVEN R TO FIND S $\dfrac{(1+I)^{**N}-1}{I}$	PRESENT WORTH FACTOR GIVEN R TO FIND P $\dfrac{(1+I)^{**N}-1}{I(1+I)^{**N}}$	N PERIODS
1	1.1000000	.9090909	1.0000000	1.1000000	1.0000000	.9090909	1
2	1.2100000	.8264463	.4761905	.5761905	2.1000000	1.7355372	2
3	1.3310000	.7513148	.3021148	.4021148	3.3100000	2.4868520	3
4	1.4641000	.6830135	.2154708	.3154708	4.6410000	3.1698654	4
5	1.6105100	.6209213	.1637975	.2637975	6.1051000	3.7907868	5
6	1.7715610	.5644739	.1296074	.2296074	7.7156100	4.3552607	6
7	1.9487171	.5131581	.1054055	.2054055	9.4871710	4.8684188	7
8	2.1435888	.4665074	.0874440	.1874440	11.4358881	5.3349262	8
9	2.3579477	.4240976	.0736405	.1736405	13.5794769	5.7590238	9
10	2.5937425	.3855433	.0627454	.1627454	15.9374246	6.1445671	10
11	2.8531167	.3504939	.0539631	.1539631	18.5311671	6.4950610	11
12	3.1384284	.3186308	.0467633	.1467633	21.3842838	6.8136918	12
13	3.4522712	.2896644	.0407785	.1407785	24.5227121	7.1033562	13
14	3.7974983	.2633313	.0357462	.1357462	27.9749834	7.3666875	14
15	4.1772482	.2393920	.0314738	.1314738	31.7724817	7.6060795	15

n							n
16	4.5949730	.2176291	.0278166	.1278166	35.9497299	7.8237086	16
17	5.0544703	.1978447	.0246641	.1246641	40.5447028	8.0215533	17
18	5.5599173	.1798588	.0219302	.1219302	45.5991731	8.2014121	18
19	6.1159090	.1635080	.0195469	.1195469	51.1590904	8.3649201	19
20	6.7274999	.1486436	.0174596	.1174596	57.2749995	8.5135637	20
21	7.4002499	.1351306	.0156244	.1156244	64.0024994	8.6486943	21
22	8.1402749	.1228460	.0140051	.1140051	71.4027494	8.7715403	22
23	8.9543024	.1116782	.0125718	.1125718	79.5430243	8.8832184	23
24	9.8497327	.1015256	.0112998	.1112998	88.4973268	8.9847440	24
25	10.8347059	.0922960	.0101681	.1101681	98.3470594	9.0770400	25
26	11.9181765	.0839055	.0091590	.1091590	109.1817654	9.1609455	26
27	13.1099942	.0762777	.0082576	.1082576	121.0999419	9.2372232	27
28	14.4209936	.0693433	.0074510	.1074510	134.2099361	9.3055665	28
29	15.8630930	.0630394	.0067281	.1067281	148.6309297	9.3696059	29
30	17.4494023	.0573086	.0060792	.1060792	164.4940227	9.4269145	30
31	19.1943425	.0520987	.0054962	.1054962	181.9434250	9.4790132	31
32	21.1137767	.0473624	.0049717	.1049717	201.1377675	9.5263756	32
33	23.2251544	.0430568	.0044994	.1044994	222.2515442	9.5694324	33
34	25.5476699	.0391425	.0040737	.1040737	245.4766986	9.6085749	34
35	28.1024368	.0355841	.0036897	.1036897	271.0243685	9.6441590	35
36	30.9126805	.0323492	.0033431	.1033431	299.1268053	9.6765082	36
37	34.0039486	.0294083	.0030299	.1030299	330.0394859	9.7059165	37
38	37.4043434	.0267349	.0027469	.1027469	364.0434344	9.7326514	38
39	41.1447778	.0243044	.0024910	.1024910	401.4477779	9.7569558	39
40	45.2592556	.0220949	.0022594	.1022594	442.5925557	9.7790507	40

NOTE- **N IS EXPONENT N

11.00 PERCENT COMPOUND INTEREST FACTORS

	------SINGLE PAYMENT------		SINKING FUND FACTOR	------UNIFORM ANNUAL SERIES------			
	COMPOUND AMOUNT FACTOR GIVEN P TO FIND S	PRESENT WORTH FACTOR GIVEN S TO FIND P	GIVEN S TO FIND R	CAPITAL RECOVERY FACTOR GIVEN P TO FIND R	COMPOUND AMOUNT FACTOR GIVEN R TO FIND S	PRESENT WORTH FACTOR GIVEN R TO FIND P	N PERIODS
N PERIODS	$(1 + I)^{**}N$	$\dfrac{1}{(1+I)^{**}N}$	$\dfrac{I}{(1+I)^{**}N - 1}$	$\dfrac{I(1+I)^{**}N}{(1+I)^{**}N - 1}$	$\dfrac{(1+I)^{**}N - 1}{I}$	$\dfrac{(1+I)^{**}N - 1}{I(1+I)^{**}N}$	
1	1.1100000	.9009009	1.0000000	1.1100000	1.0000000	.9009009	1
2	1.2321000	.8116224	.4739336	.5839336	2.1100000	1.7125233	2
3	1.3676310	.7311914	.2992131	.4092131	3.3421000	2.4437147	3
4	1.5180704	.6587310	.2123264	.3223264	4.7097310	3.1024457	4
5	1.6850582	.5934513	.1605703	.2705703	6.2278014	3.6958970	5
6	1.8704146	.5346408	.1263766	.2363766	7.9128596	4.2305379	6
7	2.0761602	.4816584	.1022153	.2122153	9.7832741	4.7121963	7
8	2.3045378	.4339265	.0843211	.1943211	11.8594343	5.1461228	8
9	2.5580369	.3909248	.0706017	.1806017	14.1639720	5.5370475	9
10	2.8394210	.3521845	.0598014	.1698014	16.7220090	5.8892320	10
11	3.1517573	.3172833	.0511210	.1611210	19.5614300	6.2065153	11
12	3.4984506	.2858408	.0440273	.1540273	22.7131872	6.4923561	12
13	3.8832802	.2575143	.0381510	.1481510	26.2116378	6.7498704	13
14	4.3104410	.2319948	.0332282	.1432282	30.0949180	6.9818652	14
15	4.7845895	.2090043	.0290652	.1390652	34.4053590	7.1908696	15

N							N
16	5.3108943	.1882922	.0255167	.1355167	39.1899485	7.3791618	16
17	5.8950927	.1696326	.0224715	.1324715	44.5008428	7.5487944	17
18	6.5435529	.1528222	.0198429	.1298429	50.3959355	7.7016166	18
19	7.2633437	.1376776	.0175625	.1275625	56.9394884	7.8392942	19
20	8.0623115	.1240339	.0155756	.1255756	64.2028321	7.9633281	20
21	8.9491658	.1117423	.0138379	.1238379	72.2651437	8.0750704	21
22	9.9335740	.1006687	.0123131	.1223131	81.2143095	8.1757391	22
23	11.0262672	.0906925	.0109712	.1209712	91.1478835	8.2664316	23
24	12.2391566	.0817050	.0097872	.1197872	102.1741507	8.3481366	24
25	13.5854638	.0736081	.0087402	.1187402	114.4133073	8.4217447	25
26	15.0798648	.0663136	.0078126	.1178126	127.9987711	8.4480583	26
27	16.7386500	.0597420	.0069892	.1169892	143.0786359	8.5478002	27
28	18.5799014	.0538216	.0062571	.1162571	159.8172859	8.6016218	28
29	20.6236906	.0484879	.0056055	.1156055	178.3971873	8.6501098	29
30	22.8922966	.0436828	.0050246	.1150246	199.0208779	8.6937926	30
31	25.4104492	.0393539	.0045063	.1145063	221.9131745	8.7331465	31
32	28.2055986	.0354540	.0040433	.1140433	247.3236237	8.7686004	32
33	31.3082145	.0319405	.0036294	.1136294	275.5292223	8.8005409	33
34	34.7521180	.0287752	.0032591	.1132591	306.8374368	8.8293161	34
35	38.5748510	.0259236	.0029275	.1129275	341.5895548	8.8552398	35
36	42.8180846	.0233546	.0026304	.1126304	380.1644058	8.8785944	36
37	47.5280740	.0210402	.0023642	.1123642	422.9824905	8.8996346	37
38	52.7561621	.0189551	.0021254	.1121254	470.5105644	8.9185897	38
39	58.5593399	.0170707	.0019111	.1119111	523.2667265	8.9356664	39
40	65.0008673	.0153844	.0017187	.1117187	581.8260664	8.9510508	40

NOTE- **N IS EXPONENT N

12.00 PERCENT COMPOUND INTEREST FACTORS

	SINGLE PAYMENT			UNIFORM ANNUAL SERIES			
N PERIODS	COMPOUND AMOUNT FACTOR GIVEN P TO FIND S $(1 + I)^{**N}$	PRESENT WORTH FACTOR GIVEN S TO FIND P $\dfrac{1}{(1 + I)^{**N}}$	SINKING FUND FACTOR GIVEN S TO FIND R $\dfrac{I}{(1 + I)^{**N} - 1}$	CAPITAL RECOVERY FACTOR GIVEN P TO FIND R $\dfrac{I(1 + I)^{**N}}{(1 + I)^{**N} - 1}$	COMPOUND AMOUNT FACTOR GIVEN R TO FIND S $\dfrac{(1 + I)^{**N} - 1}{I}$	PRESENT WORTH FACTOR GIVEN R TO FIND P $\dfrac{(1 + I)^{**N} - 1}{I(1 + I)^{**N}}$	N PERIODS
1	1.1200000	.8928571	1.0000000	1.1200000	1.0000000	.8928571	1
2	1.2544000	.7971939	.4716981	.5916981	2.1200000	1.6900510	2
3	1.4049280	.7117802	.2963490	.4163490	3.3744000	2.4018313	3
4	1.5735194	.6355181	.2092344	.3292344	4.7793280	3.0373493	4
5	1.7623417	.5674269	.1574097	.2774097	6.3528474	3.6047762	5
6	1.9738227	.5066311	.1232257	.2432257	8.1151890	4.1114073	6
7	2.2106814	.4523492	.0991177	.2191177	10.0890117	4.5637565	7
8	2.4759632	.4038832	.0813028	.2013028	12.2996931	4.9676398	8
9	2.7730788	.3606100	.0676789	.1876789	14.7756563	5.3282498	9
10	3.1058482	.3219732	.0569842	.1769842	17.5487351	5.6502230	10
11	3.4785500	.2874761	.0484154	.1684154	20.6545833	5.9376991	11
12	3.8959760	.2566731	.0414368	.1614368	24.1331333	6.1943742	12
13	4.3634931	.2291742	.0356772	.1556772	28.0291093	6.4235484	13
14	4.8871123	.2046198	.0308712	.1508712	32.3926024	6.6281682	14
15	5.4735658	.1826963	.0268242	.1468242	37.2797147	6.8108645	15

294

N							N
16	6.1303937	.1631217	.0233900	.1433900	42.7532804	6.9739862	16
17	6.8660409	.1456443	.0204567	.1404567	48.8836741	7.1196305	17
18	7.6899658	.1300396	.0179373	.1379373	55.7497150	7.2496701	18
19	8.6127617	.1161068	.0157630	.1357630	63.4396808	7.3657769	19
20	9.6462931	.1036668	.0138788	.1338788	72.0524424	7.4694436	20
21	10.8038483	.0925596	.0122401	.1322401	81.6987355	7.5620032	21
22	12.1003101	.0826425	.0108105	.1308105	92.5025838	7.6446457	22
23	13.5523473	.0737880	.0095600	.1295600	104.6028939	7.7184337	23
24	15.1786289	.0658821	.0084634	.1284634	118.1552411	7.7843158	24
25	17.0000644	.0588233	.0075000	.1275000	133.3338701	7.8431391	25
26	19.0400721	.0525208	.0066519	.1266519	150.3339345	7.8956599	26
27	21.3248808	.0468936	.0059041	.1259041	169.3740066	7.9425535	27
28	23.8838665	.0418693	.0052439	.1252439	190.6988874	7.9844228	28
29	26.7499305	.0373833	.0046602	.1246602	214.5827539	8.0218060	29
30	29.9599221	.0333779	.0041437	.1241437	241.3326843	8.0551840	30
31	33.5551128	.0298017	.0036861	.1236861	271.2926065	8.0849857	31
32	37.5817263	.0266087	.0032803	.1232803	304.8477192	8.1115944	32
33	42.0915335	.0237577	.0029203	.1229203	342.4294455	8.1353521	33
34	47.1425175	.0212123	.0026006	.1226006	384.5209790	8.1565644	34
35	52.7996196	.0189395	.0023166	.1223166	431.6634965	8.1755039	35
36	59.1355739	.0169103	.0020641	.1220641	484.4631161	8.1924142	36
37	66.2318428	.0150985	.0018396	.1218396	543.5986900	8.2075127	37
38	74.1796639	.0134808	.0016398	.1216398	609.8305328	8.2209935	38
39	83.0812236	.0120364	.0014620	.1214620	684.0101967	8.2330299	39
40	93.0509704	.0107468	.0013036	.1213036	767.0914203	8.2437767	40

NOTE- **N IS EXPONENT N

13.00 PERCENT COMPOUND INTEREST FACTORS

N PERIODS	SINGLE PAYMENT COMPOUND AMOUNT FACTOR GIVEN P TO FIND S $(1 + I)^{**N}$	SINGLE PAYMENT PRESENT WORTH FACTOR GIVEN S TO FIND P $\dfrac{1}{(1+I)^{**N}}$	UNIFORM ANNUAL SERIES SINKING FUND FACTOR GIVEN S TO FIND R $\dfrac{I}{(1+I)^{**N}-1}$	UNIFORM ANNUAL SERIES CAPITAL RECOVERY FACTOR GIVEN P TO FIND R $\dfrac{I(1+I)^{**N}}{(1+I)^{**N}-1}$	UNIFORM ANNUAL SERIES COMPOUND AMOUNT FACTOR GIVEN R TO FIND S $\dfrac{(1+I)^{**N}-1}{I}$	UNIFORM ANNUAL SERIES PRESENT WORTH FACTOR GIVEN R TO FIND P $\dfrac{(1+I)^{**N}-1}{I(1+I)^{**N}}$	N PERIODS
1	1.1300000	.8849558	1.0000000	1.1300000	1.0000000	.8849558	1
2	1.2769000	.7831467	.4694836	.5994836	2.1300000	1.6681024	2
3	1.4428970	.6930502	.2935220	.4235220	3.4069000	2.3611526	3
4	1.6304736	.6133187	.2061942	.3361942	4.8497970	2.9744713	4
5	1.8424352	.5427599	.1543145	.2843145	6.4802706	3.5172313	5
6	2.0819518	.4803185	.1201532	.2501532	8.3227058	3.9975498	6
7	2.3526055	.4250606	.0961108	.2261108	10.4046575	4.4226104	7
8	2.6584442	.3761599	.0783867	.2083867	12.7572630	4.7987703	8
9	3.0040419	.3328848	.0648689	.1948689	15.4157072	5.1316551	9
10	3.3945674	.2945883	.0542896	.1842896	18.4197492	5.4262435	10
11	3.8358612	.2606977	.0458415	.1758415	21.8143165	5.6869411	11
12	4.3345231	.2307059	.0389861	.1689861	25.6501777	5.9176470	12
13	4.8980111	.2041645	.0333503	.1633503	29.9847008	6.1218115	13
14	5.5347525	.1806766	.0286675	.1586675	34.8827119	6.3024881	14
15	6.2542704	.1598908	.0247418	.1547418	40.4174644	6.4623788	15

N							N
16	7.0673255	.1414962	.0214262	.1514262	46.6717348	6.6038751	16
17	7.9860778	.1252179	.0186084	.1486084	53.7390603	6.7290930	17
18	9.0242680	.1108123	.0162009	.1462009	61.7251382	6.8399053	18
19	10.1974228	.0980640	.0141344	.1441344	70.7494062	6.9379693	19
20	11.5230878	.0867823	.0123538	.1423538	80.9468290	7.0247516	20
21	13.0210892	.0767985	.0108143	.1408143	92.4699167	7.1015501	21
22	14.7138308	.0679633	.0094795	.1394795	105.4910059	7.1695133	22
23	16.6266288	.0601445	.0083191	.1383191	120.2048367	7.2296578	23
24	18.7880905	.0532252	.0073083	.1373083	136.8314654	7.2828830	24
25	21.2305423	.0471020	.0064259	.1364259	155.6195559	7.3299850	25
26	23.9905128	.0416831	.0056545	.1356545	176.8500982	7.3716681	26
27	27.1092794	.0368877	.0049791	.1349791	200.8406110	7.4085559	27
28	30.6334858	.0326440	.0043869	.1343869	227.9498904	7.4411999	28
29	34.6158389	.0288885	.0038672	.1338672	258.5833762	7.4700884	29
30	39.1158980	.0255651	.0034107	.1334107	293.1992151	7.4956534	30
31	44.2009647	.0226239	.0030092	.1330092	332.3151130	7.5182774	31
32	49.9470901	.0200212	.0026559	.1326559	376.5160777	7.5382986	32
33	56.4402118	.0177179	.0023449	.1323449	426.4631678	7.5560164	33
34	63.7774394	.0156795	.0020708	.1320708	482.9033796	7.5716960	34
35	72.0685065	.0138757	.0018292	.1318292	546.6808190	7.5855716	35
36	81.4374123	.0122794	.0016162	.1316162	618.7493254	7.5978510	36
37	92.0242759	.0108667	.0014282	.1314282	700.1867377	7.6087177	37
38	103.9874318	.0096165	.0012623	.1312623	792.2110137	7.6183343	38
39	117.5057979	.0085102	.0011158	.1311158	896.1984454	7.6268445	39
40	132.7815516	.0075312	.0009865	.1309865	1013.7042433	7.6343756	40

NOTE- **N IS EXPONENT N

14.00 PERCENT COMPOUND INTEREST FACTORS

	SINGLE PAYMENT		UNIFORM ANNUAL SERIES				
N PERIODS	COMPOUND AMOUNT FACTOR GIVEN P TO FIND S	PRESENT WORTH FACTOR GIVEN S TO FIND P	SINKING FUND FACTOR GIVEN S TO FIND R	CAPITAL RECOVERY FACTOR GIVEN P TO FIND R	COMPOUND AMOUNT FACTOR GIVEN R TO FIND S	PRESENT WORTH FACTOR GIVEN R TO FIND P	N PERIODS
	$(1+I)^{**}N$	$\dfrac{1}{(1+I)^{**}N}$	$\dfrac{I}{(1+I)^{**}N-1}$	$\dfrac{I(1+I)^{**}N}{(1+I)^{**}N-1}$	$\dfrac{(1+I)^{**}N-1}{I}$	$\dfrac{(1+I)^{**}N-1}{I(1+I)^{**}N}$	
1	1.1400000	.8771930	1.0000000	1.1400000	1.0000000	.8771930	1
2	1.2996000	.7694675	.4672897	.6072897	2.1400000	1.6466605	2
3	1.4815440	.6749715	.2907315	.4307315	3.4396000	2.3216320	3
4	1.6889602	.5920803	.2032048	.3432048	4.9211440	2.9137123	4
5	1.9254146	.5193687	.1512835	.2912835	6.6101042	3.4330810	5
6	2.1949726	.4555865	.1171575	.2571575	8.5355187	3.8886675	6
7	2.5022688	.3996373	.0931924	.2331924	10.7304914	4.2883048	7
8	2.8525864	.3505079	.0755700	.2155700	13.2327602	4.6388639	8
9	3.2519485	.3075079	.0621684	.2021684	16.0853466	4.9463718	9
10	3.7072213	.2697438	.0517135	.1917135	19.3372951	5.2161156	10
11	4.2262323	.2366174	.0433943	.1833943	23.0445164	5.4527330	11
12	4.8179048	.2075591	.0366693	.1766693	27.2707487	5.6602921	12
13	5.4924115	.1820694	.0311637	.1711637	32.0886535	5.8423615	13
14	6.2613491	.1597100	.0266091	.1666091	37.5810650	6.0020715	14
15	7.1379380	.1400965	.0228090	.1628090	43.8424141	6.1421680	15

n							n
16	8.1372493	.1228917	.0196154	.1596154	50.9803521	6.2650596	16
17	9.2764642	.1077997	.0169154	.1569154	59.1176014	6.3728593	17
18	10.5751692	.0945611	.0146212	.1546212	68.3940656	6.4674205	18
19	12.0556929	.0829484	.0126632	.1526632	78.9692348	6.5503688	19
20	13.7434899	.0727617	.0109860	.1509860	91.0249277	6.6231306	20
21	15.6675785	.0638261	.0095449	.1495449	104.7684175	6.6895566	21
22	17.8610394	.0559878	.0083032	.1483032	120.4359960	6.7429444	22
23	20.3615850	.0491121	.0072308	.1472308	138.2970354	6.7920565	23
24	23.2122069	.0430808	.0063028	.1463028	158.6586204	6.8351373	24
25	26.4619158	.0377902	.0054984	.1454984	181.8708272	6.8729274	25
26	30.1665840	.0331493	.0048000	.1448000	208.3327430	6.9060767	26
27	34.3899058	.0290783	.0041929	.1441929	238.4993271	6.9351550	27
28	39.2044926	.0255073	.0036645	.1436645	272.8892329	6.9606623	28
29	44.6931216	.0223748	.0032042	.1432042	312.0937255	6.9830371	29
30	50.9501586	.0196270	.0028028	.1428028	356.7868470	7.0026641	30
31	58.0831808	.0172167	.0024526	.1424526	407.7370056	7.0198808	31
32	66.2148261	.0151024	.0021468	.1421468	465.8201864	7.0349832	32
33	75.4849017	.0132477	.0018796	.1418796	532.0350125	7.0482308	33
34	86.0527880	.0116208	.0016460	.1416460	607.5199142	7.0598516	34
35	98.1001783	.0101937	.0014418	.1414418	693.5727022	7.0700453	35
36	111.8342033	.0089418	.0012631	.1412631	791.6728805	7.0789871	36
37	127.4909917	.0078437	.0011068	.1411068	903.5070838	7.0868308	37
38	145.3397306	.0068804	.0009699	.1409699	1030.9980755	7.0937112	38
39	165.6872929	.0060355	.0008501	.1408501	1176.3378061	7.0997467	39
40	188.8835139	.0052943	.0007451	.1407451	1342.0250990	7.1050409	40

NOTE— **N IS EXPONENT N

299

15.00 PERCENT COMPOUND INTEREST FACTORS

N PERIODS	SINGLE PAYMENT COMPOUND AMOUNT FACTOR GIVEN P TO FIND S $(1+I)^{**N}$	PRESENT WORTH FACTOR GIVEN S TO FIND P $\dfrac{1}{(1+I)^{**N}}$	SINKING FUND FACTOR GIVEN S TO FIND R $\dfrac{I}{(1+I)^{**N}-1}$	UNIFORM ANNUAL SERIES CAPITAL RECOVERY FACTOR GIVEN P TO FIND R $\dfrac{I(1+I)^{**N}}{(1+I)^{**N}-1}$	COMPOUND AMOUNT FACTOR GIVEN R TO FIND S $\dfrac{(1+I)^{**N}-1}{I}$	PRESENT WORTH FACTOR GIVEN R TO FIND P $\dfrac{(1+I)^{**N}-1}{I(1+I)^{**N}}$	N PERIODS
1	1.1500000	.8695652	1.0000000	1.1500000	1.0000000	.8695652	1
2	1.3225000	.7561437	.4651163	.6151163	2.1500000	1.6257089	2
3	1.5208750	.6575162	.2879770	.4379770	3.4725000	2.2832251	3
4	1.7490063	.5717532	.2002654	.3502654	4.9933750	2.8549784	4
5	2.0113572	.4971767	.1483156	.2983156	6.7423813	3.3521551	5
6	2.3130608	.4323276	.1142369	.2642369	8.7537384	3.7844827	6
7	2.6600199	.3759370	.0903604	.2403604	11.0667992	4.1604197	7
8	3.0590229	.3269018	.0728501	.2228501	13.7268191	4.4873215	8
9	3.5178763	.2842624	.0595740	.2095740	16.7858419	4.7715839	9
10	4.0455577	.2471847	.0492521	.1992521	20.3037182	5.0187686	10
11	4.6523914	.2149432	.0410690	.1910690	24.3492760	5.2337118	11
12	5.3502501	.1869072	.0344808	.1844808	29.0016674	5.4206190	12
13	6.1527876	.1625280	.0291105	.1791105	34.3519175	5.5831470	13
14	7.0757058	.1413287	.0246885	.1746885	40.5047051	5.7244756	14
15	8.1370616	.1228945	.0210171	.1710171	47.5804109	5.8473701	15
16	9.3576209	.1068648	.0179477	.1679477	55.7174725	5.9542349	16

15.00%

n							n
17	10.7612640	.0929259	.0153669	.1653669	65.0750934	6.0471608	17
18	12.3754536	.0808051	.0131863	.1631863	75.8363574	6.1279659	18
19	14.2317716	.0702653	.0113364	.1613364	88.2118110	6.1982312	19
20	16.3665374	.0611003	.0097615	.1597615	102.4435826	6.2593315	20
21	18.8215180	.0531307	.0084168	.1584168	118.8101200	6.3124622	21
22	21.6447457	.0462006	.0072658	.1572658	137.6316380	6.3586627	22
23	24.8914576	.0401744	.0062784	.1562784	159.2763837	6.3988372	23
24	28.6251762	.0349343	.0054298	.1554298	184.1678413	6.4337714	24
25	32.9189526	.0303776	.0046994	.1546994	212.7930175	6.4641491	25
26	37.8567955	.0264153	.0040698	.1540698	245.7119701	6.4905644	26
27	43.5353148	.0229699	.0035265	.1535265	283.5687656	6.5135343	27
28	50.0656121	.0199738	.0030571	.1530571	327.1040804	6.5335081	28
29	57.5754539	.0173685	.0026513	.1526513	377.1696925	6.5508766	29
30	66.2117720	.0151031	.0023002	.1523002	434.7451464	6.5659796	30
31	76.1435378	.0131331	.0019962	.1519962	500.9569183	6.5791127	31
32	87.5650684	.0114201	.0017328	.1517328	577.1004561	6.5905328	32
33	100.6998287	.0099305	.0015045	.1515045	664.6655245	6.6004633	33
34	115.8048030	.0086352	.0013066	.1513066	765.3653532	6.6090985	34
35	133.1755234	.0075089	.0011349	.1511349	881.1701561	6.6166074	35
36	153.1518519	.0065295	.0009859	.1509859	1014.3456796	6.6231369	36
37	176.1246297	.0056778	.0008565	.1508565	1167.4975315	6.6288147	37
38	202.5432242	.0049372	.0007443	.1507443	1343.6221612	6.6337519	38
39	232.9248228	.0042932	.0006468	.1506468	1546.1654854	6.6380451	39
40	267.8635462	.0037332	.0005621	.1505621	1779.0903082	6.6417784	40

NOTE- **N IS EXPONENT N

16.00 PERCENT COMPOUND INTEREST FACTORS

	SINGLE PAYMENT		UNIFORM ANNUAL SERIES				
N PERIODS	COMPOUND AMOUNT FACTOR GIVEN P TO FIND S $(1 + I)^{**N}$	PRESENT WORTH FACTOR GIVEN S TO FIND P $\dfrac{1}{(1 + I)^{**N}}$	SINKING FUND FACTOR GIVEN S TO FIND R $\dfrac{I}{(1 + I)^{**N} - 1}$	CAPITAL RECOVERY FACTOR GIVEN P TO FIND R $\dfrac{I(1 + I)^{**N}}{(1 + I)^{**N} - 1}$	COMPOUND AMOUNT FACTOR GIVEN R TO FIND S $\dfrac{(1 + I)^{**N} - 1}{I}$	PRESENT WORTH FACTOR GIVEN R TO FIND P $\dfrac{(1 + I)^{**N} - 1}{I(1 + I)^{**N}}$	N PERIODS
1	1.1600000	.8620690	1.0000000	1.1600000	1.0000000	.8620690	1
2	1.3456000	.7431629	.4629630	.6229630	2.1600000	1.6052319	2
3	1.5608960	.6406577	.2852579	.4452579	3.5056000	2.2458895	3
4	1.8106394	.5522911	.1973751	.3573751	5.0664960	2.7981806	4
5	2.1003417	.4761130	.1454094	.3054094	6.8771354	3.2742937	5
6	2.4363963	.4104423	.1113899	.2713899	8.9774770	3.6847359	6
7	2.8262197	.3538295	.0876127	.2476127	11.4138733	4.0385654	7
8	3.2784149	.3050255	.0702243	.2302243	14.2400931	4.3435909	8
9	3.8029613	.2629530	.0570825	.2170825	17.5185080	4.6065439	9
10	4.4114351	.2266836	.0469011	.2069011	21.3214692	4.8332275	10
11	5.1172647	.1954169	.0388608	.1988608	25.7329043	5.0286444	11
12	5.9360270	.1684628	.0324147	.1924147	30.8501690	5.1971072	12
13	6.8857914	.1452266	.0271841	.1871841	36.7861961	5.3423338	13
14	7.9875180	.1251953	.0228980	.1828980	43.6719874	5.4675291	14
15	9.2655209	.1079270	.0193575	.1793575	51.6595054	5.5754562	15

N							N
16	10.7480042	.0930405	.0164136	.1764136	60.9250263	5.6684967	16
17	12.4676849	.0802074	.0139522	.1739522	71.6730305	5.7487040	17
18	14.4625145	.0691443	.0118849	.1718849	84.1407154	5.8178483	18
19	16.7765168	.0596071	.0101417	.1701417	98.6032298	5.8774554	19
20	19.4607595	.0513855	.0086670	.1686670	115.3797466	5.9288409	20
21	22.5744810	.0442978	.0074162	.1674162	134.8405060	5.9731387	21
22	26.1863979	.0381878	.0063526	.1663526	157.4149870	6.0113265	22
23	30.3762216	.0329205	.0054466	.1654466	183.6013849	6.0442470	23
24	35.2364170	.0283797	.0046734	.1646734	213.9776065	6.0726267	24
25	40.8742438	.0244653	.0040126	.1640126	249.2140235	6.0970920	25
26	47.4141228	.0210908	.0034472	.1634472	290.0882673	6.1181827	26
27	55.0003824	.0181817	.0029629	.1629629	337.5023901	6.1363644	27
28	63.8004436	.0156739	.0025478	.1625478	392.5027725	6.1520383	28
29	74.0085146	.0135120	.0021915	.1621915	456.3032161	6.1655503	29
30	85.8498769	.0116482	.0018857	.1618857	530.3117307	6.1771985	30
31	99.5858572	.0100416	.0016230	.1616230	616.1616076	6.1872401	31
32	115.5195944	.0086565	.0013971	.1613971	715.7474648	6.1958966	32
33	134.0027295	.0074625	.0012030	.1612030	831.2670592	6.2033592	33
34	155.4431662	.0064332	.0010360	.1610360	965.2697886	6.2097924	34
35	180.3140728	.0055459	.0008923	.1608923	1120.7129548	6.2153383	35
36	209.1643244	.0047809	.0007686	.1607686	1301.0270276	6.2201192	36
37	242.6306163	.0041215	.0006622	.1606622	1510.1913520	6.2242407	37
38	281.4515149	.0035530	.0005705	.1605705	1752.8219683	6.2277937	38
39	326.4837573	.0030629	.0004916	.1604916	2034.2734833	6.2308566	39
40	378.7211585	.0026405	.0004236	.1604236	2360.7572406	6.2334971	40

NOTE- .**N IS EXPONENT N

17.00 PERCENT COMPOUND INTEREST FACTORS

	SINGLE PAYMENT			UNIFORM ANNUAL SERIES			
N PERIODS	COMPOUND AMOUNT FACTOR GIVEN P TO FIND S $(1+I)^N$	PRESENT WORTH FACTOR GIVEN S TO FIND P $\dfrac{1}{(1+I)^N}$	SINKING FUND FACTOR GIVEN S TO FIND R $\dfrac{I}{(1+I)^N-1}$	CAPITAL RECOVERY FACTOR GIVEN P TO FIND R $\dfrac{I(1+I)^N}{(1+I)^N-1}$	COMPOUND AMOUNT FACTOR GIVEN R TO FIND S $\dfrac{(1+I)^N-1}{I}$	PRESENT WORTH FACTOR GIVEN R TO FIND P $\dfrac{(1+I)^N-1}{I(1+I)^N}$	N PERIODS
1	1.1700000	.8547009	1.0000000	1.1700000	1.0000000	.8547009	1
2	1.3689000	.7305136	.4608295	.6308295	2.1700000	1.5852144	2
3	1.6016130	.6243706	.2825737	.4525737	3.5389000	2.2095850	3
4	1.8738872	.5336500	.1945331	.3645331	5.1405130	2.7432350	4
5	2.1924480	.4561112	.1425639	.3125639	7.0144002	3.1993462	5
6	2.5651642	.3898386	.1086148	.2786148	9.2068482	3.5891848	6
7	3.0012421	.3331954	.0849472	.2549472	11.7720124	3.9223801	7
8	3.5114533	.2847824	.0676899	.2376899	14.7732546	4.2071625	8
9	4.1084003	.2434037	.0546905	.2246905	18.2847078	4.4505662	9
10	4.8068284	.2080374	.0446566	.2146566	22.3931082	4.6586036	10
11	5.6239892	.1778097	.0367648	.2067648	27.1999366	4.8364134	11
12	6.5800674	.1519741	.0304656	.2004656	32.8239258	4.9883875	12
13	7.6986788	.1298924	.0253781	.1953781	39.4039932	5.1182799	13
14	9.0074542	.1110192	.0212302	.1912302	47.1026720	5.2292991	14
15	10.5387215	.0948882	.0178221	.1878221	56.1101262	5.3241872	15

n							n
16	12.3303041	.0811010	.0150040	.1850040	66.6488477	5.4052882	16
17	14.4264558	.0693171	.0126616	.1826616	78.9791518	5.4746053	17
18	16.8789533	.0592454	.0107060	.1807060	93.4056076	5.5338507	18
19	19.7483754	.0506371	.0090675	.1790675	110.2845609	5.5844878	19
20	23.1055992	.0432796	.0076904	.1776904	130.0329363	5.6277673	20
21	27.0335510	.0369911	.0065300	.1765300	153.1385354	5.6647584	21
22	31.6292547	.0316163	.0055502	.1755502	180.1720864	5.6963747	22
23	37.0062280	.0270225	.0047214	.1747214	211.8013411	5.7233972	23
24	43.2972868	.0230961	.0040192	.1740192	248.8075691	5.7464933	24
25	50.6578255	.0197403	.0034234	.1734234	292.1048559	5.7662336	25
26	59.2696558	.0168720	.0029175	.1729175	342.7626814	5.7831056	26
27	69.3454973	.0144205	.0024874	.1724874	402.0323372	5.7975262	27
28	81.1342319	.0123253	.0021214	.1721214	471.3778345	5.8098514	28
29	94.9270513	.0105344	.0018099	.1718099	552.5120664	5.8203859	29
30	111.0646500	.0090038	.0015445	.1715445	647.4391177	5.8293896	30
31	129.9456405	.0076955	.0013184	.1713184	758.5037677	5.8370851	31
32	152.0363994	.0065774	.0011256	.1711256	888.4494082	5.8436625	32
33	177.8825873	.0056217	.0009611	.1709611	1040.4858076	5.8492842	33
34	208.1226271	.0048049	.0008208	.1708208	1218.3683949	5.8540891	34
35	243.5034738	.0041067	.0007010	.1707010	1426.4910221	5.8581958	35
36	284.8990643	.0035100	.0005988	.1705988	1669.9944958	5.8617058	36
37	333.3319052	.0030000	.0005115	.1705115	1954.8935601	5.8647058	37
38	389.9983291	.0025641	.0004370	.1704370	2288.2254653	5.8672699	38
39	456.2980451	.0021916	.0003734	.1703734	2678.2237944	5.8694615	39
40	533.8687127	.0018731	.0003190	.1703190	3134.5218395	5.8713346	40

NOTE- **N IS EXPONENT N

305

18.00 PERCENT COMPOUND INTEREST FACTORS

| | ------SINGLE PAYMENT------ | | ------UNIFORM ANNUAL SERIES------ | | | |
| | COMPOUND AMOUNT FACTOR GIVEN P TO FIND S | PRESENT WORTH FACTOR GIVEN S TO FIND P | SINKING FUND FACTOR GIVEN S TO FIND R | CAPITAL RECOVERY FACTOR GIVEN P TO FIND R | COMPOUND AMOUNT FACTOR GIVEN R TO FIND S | PRESENT WORTH FACTOR GIVEN R TO FIND P | |
N PERIODS	$(1 + I)**N$	$\dfrac{1}{(1 + I)**N}$	$\dfrac{I}{(1 + I)**N - 1}$	$\dfrac{I(1 + I)**N}{(1 + I)**N - 1}$	$\dfrac{(1 + I)**N - 1}{I}$	$\dfrac{(1 + I)**N - 1}{I(1 + I)**N}$	N PERIODS
1	1.1800000	.8474576	1.0000000	1.1800000	1.0000000	.8474576	1
2	1.3924000	.7181844	.4587156	.6387156	2.1800000	1.5656421	2
3	1.6430320	.6086309	.2799239	.4599239	3.5724000	2.1742729	3
4	1.9387778	.5157889	.1917387	.3717387	5.2154320	2.6900618	4
5	2.2877578	.4371092	.1397778	.3197778	7.1542098	3.1271710	5
6	2.6995542	.3704315	.1059101	.2859101	9.4419675	3.4976026	6
7	3.1854739	.3139250	.0823620	.2623620	12.1415217	3.8115276	7
8	3.7588592	.2660382	.0652444	.2452444	15.3269956	4.0775658	8
9	4.4354539	.2254561	.0523948	.2323948	19.0858548	4.3030218	9
10	5.2338356	.1910645	.0425146	.2225146	23.5213086	4.4940863	10
11	6.1759260	.1619190	.0347764	.2147764	28.7551442	4.6560053	11
12	7.2875926	.1372195	.0286278	.2086278	34.9310701	4.7932249	12
13	8.5993593	.1162877	.0236862	.2036862	42.2186628	4.9095126	13
14	10.1472440	.0985489	.0196781	.1996781	50.8180221	5.0080615	14
15	11.9737479	.0835160	.0164028	.1964028	60.9652660	5.0915776	15

N							N
16	14.1290225	.0707763	.0137101	.1937101	72.9390139	5.1623539	16
17	16.6722466	.0599799	.0114853	.1914853	87.0680364	5.2223338	17
18	19.6732509	.0508304	.0096395	.1896395	103.7402830	5.2731642	18
19	23.2144361	.0430766	.0081028	.1881028	123.4135339	5.3162409	19
20	27.3930346	.0365056	.0068200	.1868200	146.6279700	5.3527465	20
21	32.3237808	.0309370	.0057464	.1857464	174.0210046	5.3836835	21
22	38.1420614	.0262178	.0048463	.1848463	206.3447855	5.4099012	22
23	45.0076324	.0222185	.0040902	.1840902	244.4868468	5.4321197	23
24	53.1090063	.0188292	.0034543	.1834543	289.4944793	5.4509489	24
25	62.6686274	.0159569	.0029188	.1829188	342.6034855	5.4669058	25
26	73.9489803	.0135228	.0024675	.1824675	405.2721129	5.4804287	26
27	87.2597968	.0114600	.0020867	.1820867	479.2210933	5.4918887	27
28	102.9665602	.0097119	.0017653	.1817653	566.4808901	5.5016006	28
29	121.5005410	.0082304	.0014938	.1814938	669.4474503	5.5098310	29
30	143.3706384	.0069749	.0012643	.1812643	790.9479913	5.5168060	30
31	169.1773534	.0059110	.0010703	.1810703	934.3186298	5.5227169	31
32	199.6292770	.0050093	.0009062	.1809062	1103.4959831	5.5277262	32
33	235.5625468	.0042452	.0007674	.1807674	1303.1252601	5.5319713	33
34	277.9638052	.0035976	.0006499	.1806499	1538.6878069	5.5355689	34
35	327.9972902	.0030488	.0005505	.1805505	1816.6516121	5.5386177	35
36	387.0368024	.0025837	.0004663	.1804663	2144.6489023	5.5412015	36
37	456.7034269	.0021896	.0003950	.1803950	2531.6857047	5.5433911	37
38	538.9100437	.0018556	.0003346	.1803346	2988.3891316	5.5452467	38
39	635.9138515	.0015725	.0002835	.1802835	3527.2991753	5.5468192	39
40	750.3783448	.0013327	.0002402	.1802402	4163.2130268	5.5481519	40

NOTE- **N IS EXPONENT N

19.00 PERCENT COMPOUND INTEREST FACTORS

	SINGLE PAYMENT			UNIFORM ANNUAL SERIES			
	COMPOUND AMOUNT FACTOR GIVEN P TO FIND S	PRESENT WORTH FACTOR GIVEN S TO FIND P	SINKING FUND FACTOR GIVEN S TO FIND R	CAPITAL RECOVERY FACTOR GIVEN P TO FIND R	COMPOUND AMOUNT FACTOR GIVEN R TO FIND S	PRESENT WORTH FACTOR GIVEN R TO FIND P	N PERIODS
N PERIODS	$(1 + I)^{**}N$	$\dfrac{1}{(1+I)^{**}N}$	$\dfrac{I}{(1+I)^{**}N - 1}$	$\dfrac{I(1+I)^{**}N}{(1+I)^{**}N - 1}$	$\dfrac{(1+I)^{**}N - 1}{I}$	$\dfrac{(1+I)^{**}N - 1}{I(1+I)^{**}N}$	
1	1.1900000	.8403361	1.0000000	1.1900000	1.0000000	.8403361	1
2	1.4161000	.7061648	.4566210	.6466210	2.1900000	1.5465010	2
3	1.6851590	.5934158	.2773079	.4673079	3.6061000	2.1399168	3
4	2.0053392	.4986688	.1889909	.3789909	5.2912590	2.6385855	4
5	2.3863537	.4190494	.1370502	.3270502	7.2965982	3.0576349	5
6	2.8397609	.3521423	.1032743	.2932743	9.6829519	3.4097772	6
7	3.3793154	.2959179	.0798549	.2698549	12.5227127	3.7056951	7
8	4.0213853	.2486705	.0628851	.2528851	15.9020281	3.9543657	8
9	4.7854486	.2089668	.0501922	.2401922	19.9234135	4.1633325	9
10	5.6946838	.1756024	.0404713	.2304713	24.7088621	4.3389349	10
11	6.7766737	.1475650	.0328909	.2228909	30.4035458	4.4864999	11
12	8.0642417	.1240042	.0268960	.2168960	37.1802196	4.6105041	12
13	9.5964476	.1042052	.0221022	.2121022	45.2444613	4.7147093	13
14	11.4197727	.0875674	.0182346	.2082346	54.8409089	4.8022768	14
15	13.5895295	.0735861	.0150919	.2050919	66.2606816	4.8758628	15

N							N
16	4.9376998	79.8502111	.2025234	.0125234	.0618370	16.1715401	16
17	4.9896637	96.0217512	.2004143	.0104143	.0519639	19.2441327	17
18	5.0333309	115.2658839	.1986756	.0086756	.0436671	22.9005180	18
19	5.0700259	138.1664019	.1972376	.0072376	.0366951	27.2516164	19
20	5.1008621	165.4180183	.1960453	.0060453	.0308362	32.4294235	20
21	5.1267749	197.8474417	.1950544	.0050544	.0259128	38.5910139	21
22	5.1485503	236.4384557	.1942294	.0042294	.0217754	45.9233066	22
23	5.1668490	282.3617622	.1935416	.0035416	.0182987	54.6487348	23
24	5.1822261	337.0104971	.1929673	.0029673	.0153770	65.0319944	24
25	5.1951480	402.0424915	.1924873	.0024873	.0129219	77.3880734	25
26	5.2060067	479.4305649	.1920858	.0020858	.0108587	92.0918073	26
27	5.2151317	571.5223722	.1917497	.0017497	.0091250	109.5892507	27
28	5.2227997	681.1116229	.1914682	.0014682	.0076681	130.4112084	28
29	5.2292435	811.5228813	.1912323	.0012323	.0064437	155.1893379	29
30	5.2346584	966.7121692	.1910344	.0010344	.0054149	184.6753122	30
31	5.2392087	1151.3874814	.1908685	.0008685	.0045503	219.7636215	31
32	5.2430325	1371.1511029	.1907293	.0007293	.0038238	261.5187095	32
33	5.2462458	1632.6698124	.1906125	.0006125	.0032133	311.2072644	33
34	5.2489461	1943.8770767	.1905144	.0005144	.0027002	370.3366446	34
35	5.2512152	2314.2137213	.1904321	.0004321	.0022691	440.7006071	35
36	5.2531220	2754.9143284	.1903630	.0003630	.0019068	524.4337224	36
37	5.2547244	3279.3480508	.1903049	.0003049	.0016024	624.0761296	37
38	5.2560709	3903.4241804	.1902562	.0002562	.0013465	742.6505943	38
39	5.2572024	4646.0747747	.1902152	.0002152	.0011315	883.7542072	39
40	5.2581533	5529.8289819	.1901808	.0001808	.0009509	1051.6675066	40

NOTE- **N IS EXPONENT N

20.00 PERCENT COMPOUND INTEREST FACTORS

N PERIODS	SINGLE PAYMENT COMPOUND AMOUNT FACTOR GIVEN P TO FIND S $(1 + I)**N$	SINGLE PAYMENT PRESENT WORTH FACTOR GIVEN S TO FIND P $\dfrac{1}{(1 + I)**N}$	SINKING FUND FACTOR GIVEN S TO FIND R $\dfrac{I}{(1 + I)**N - 1}$	UNIFORM ANNUAL SERIES CAPITAL RECOVERY FACTOR GIVEN P TO FIND R $\dfrac{I(1 + I)**N}{(1 + I)**N - 1}$	UNIFORM ANNUAL SERIES COMPOUND AMOUNT FACTOR GIVEN R TO FIND S $\dfrac{(1 + I)**N - 1}{I}$	UNIFORM ANNUAL SERIES PRESENT WORTH FACTOR GIVEN R TO FIND P $\dfrac{(1 + I)**N - 1}{I(1 + I)**N}$	N PERIODS
1	1.2000000	.8333333	1.0000000	1.2000000	1.0000000	.8333333	1
2	1.4400000	.6944444	.4545455	.6545455	2.2000000	1.5277778	2
3	1.7280000	.5787037	.2747253	.4747253	3.6400000	2.1064815	3
4	2.0736000	.4822531	.1862891	.3862891	5.3680000	2.5887346	4
5	2.4883200	.4018776	.1343797	.3343797	7.4416000	2.9906121	5
6	2.9859840	.3348980	.1007057	.3007057	9.9299200	3.3251101	6
7	3.5831808	.2790816	.0774239	.2774239	12.9159040	3.6045918	7
8	4.2998170	.2325680	.0606094	.2606094	16.4990848	3.8371598	8
9	5.1597804	.1938067	.0480795	.2480795	20.7989018	4.0309665	9
10	6.1917364	.1615056	.0385228	.2385228	25.9586821	4.1924721	10
11	7.4300837	.1345880	.0311038	.2311038	32.1504185	4.3270601	11
12	8.9161004	.1121567	.0252650	.2252650	39.5805022	4.4392167	12
13	10.6993205	.0934639	.0206200	.2206200	48.4966027	4.5326806	13
14	12.8391846	.0778866	.0168931	.2168931	59.1959232	4.6105672	14
15	15.4070216	.0649055	.0138821	.2138821	72.0351079	4.6754726	15

n							n
16	4.7295605	87.4421294	.2114361	.0114361	.0540879	18.4884259	16
17	4.7746338	105.9305553	.2094401	.0094401	.0450732	22.1861111	17
18	4.8121948	128.1166664	.2078054	.0078054	.0375610	26.6233333	18
19	4.8434957	154.7399997	.2064625	.0064625	.0313009	31.9479999	19
20	4.8695797	186.6879996	.2053565	.0053565	.0260841	38.3375999	20
21	4.8913164	225.0255995	.2044439	.0044439	.0217367	46.0051199	21
22	4.9094304	271.0307195	.2036896	.0036896	.0181139	55.2061439	22
23	4.9245253	326.2368633	.2030653	.0030653	.0150949	66.2473727	23
24	4.9371044	392.4842360	.2025479	.0025479	.0125791	79.4968472	24
25	4.9475870	471.9810832	.2021187	.0021187	.0104826	95.3962166	25
26	4.9563225	567.3772999	.2017625	.0017625	.0087355	114.4754600	26
27	4.9636021	681.8527598	.2014666	.0014666	.0072796	137.3705520	27
28	4.9696684	819.2233118	.2012207	.0012207	.0060663	164.8446624	28
29	4.9747237	984.0679742	.2010162	.0010162	.0050553	197.8135948	29
30	4.9789364	1181.8815690	.2008461	.0008461	.0042127	237.3763138	30
31	4.9824470	1419.2578828	.2007046	.0007046	.0035106	284.8515766	31
32	4.9853725	1704.1094594	.2005868	.0005868	.0029255	341.8218919	32
33	4.9878104	2045.9313512	.2004888	.0004888	.0024379	410.1862702	33
34	4.9898420	2456.1176215	.2004071	.0004071	.0020316	492.2235243	34
35	4.9915350	2948.3411458	.2003392	.0003392	.0016930	590.6682292	35
36	4.9929458	3539.0093749	.2002826	.0002826	.0014108	708.8018750	36
37	4.9941215	4247.8112499	.2002354	.0002354	.0011757	850.5622500	37
38	4.9951013	5098.3734999	.2001961	.0001961	.0009797	1020.6747000	38
39	4.9959177	6119.0481999	.2001634	.0001634	.0008165	1224.8096400	39
40	4.9965981	7343.8578398	.2001362	.0001362	.0006804	1469.7715680	40

NOTE— **N IS EXPONENT N

21.00 PERCENT COMPOUND INTEREST FACTORS

N PERIODS	SINGLE PAYMENT — COMPOUND AMOUNT FACTOR GIVEN P TO FIND S $(1+I)^{**}N$	SINGLE PAYMENT — PRESENT WORTH FACTOR GIVEN S TO FIND P $\dfrac{1}{(1+I)^{**}N}$	UNIFORM ANNUAL SERIES — SINKING FUND FACTOR GIVEN S TO FIND R $\dfrac{I}{(1+I)^{**}N-1}$	UNIFORM ANNUAL SERIES — CAPITAL RECOVERY FACTOR GIVEN P TO FIND R $\dfrac{I(1+I)^{**}N}{(1+I)^{**}N-1}$	UNIFORM ANNUAL SERIES — COMPOUND AMOUNT FACTOR GIVEN R TO FIND S $\dfrac{(1+I)^{**}N-1}{I}$	UNIFORM ANNUAL SERIES — PRESENT WORTH FACTOR GIVEN R TO FIND P $\dfrac{(1+I)^{**}N-1}{I(1+I)^{**}N}$	N PERIODS
1	1.2100000	.8264463	1.0000000	1.2100000	1.0000000	.8264463	1
2	1.4641000	.6830135	.4524887	.6624887	2.2100000	1.5094597	2
3	1.7715610	.5644739	.2721755	.4821755	3.6741000	2.0739337	3
4	2.1435888	.4665074	.1836324	.3936324	5.4456610	2.5404410	4
5	2.5937425	.3855433	.1317653	.3417653	7.5892498	2.9259843	5
6	3.1384284	.3186308	.0982030	.3082030	10.1829923	3.2446152	6
7	3.7974983	.2633313	.0750671	.2850671	13.3214206	3.5079464	7
8	4.5949730	.2176291	.0584149	.2684149	17.1189190	3.7255755	8
9	5.5599173	.1798588	.0460535	.2560535	21.7138920	3.9054343	9
10	6.7274999	.1486436	.0366652	.2466652	27.2738093	4.0540780	10
11	8.1402749	.1228460	.0294106	.2394106	34.0013092	4.1769239	11
12	9.8497327	.1015256	.0237295	.2337295	42.1415842	4.2784495	12
13	11.9181765	.0839055	.0192340	.2292340	51.9913168	4.3623550	13
14	14.4209936	.0693433	.0156471	.2256471	63.9094934	4.4316983	14
15	17.4494023	.0573086	.0127664	.2227664	78.3304870	4.4890069	15

N							N
16	21.1137767	.0473624	.0104406	.2204406	95.7798893	4.5363693	16
17	25.5476699	.0391425	.0085548	.2185548	116.8936660	4.5755118	17
18	30.9126805	.0323492	.0070204	.2170204	142.4413359	4.6078610	18
19	37.4043434	.0267349	.0057685	.2157685	173.3540164	4.6345959	19
20	45.2592556	.0220949	.0047448	.2147448	210.7583598	4.6566908	20
21	54.7636992	.0182603	.0039060	.2139060	256.0176154	4.6749511	21
22	66.2640761	.0150911	.0032177	.2132177	310.7813147	4.6900422	22
23	80.1795321	.0124720	.0026522	.2126522	377.0453907	4.7025142	23
24	97.0172238	.0103074	.0021871	.2121871	457.2249228	4.7128217	24
25	117.3908529	.0085186	.0018043	.2118043	554.2421566	4.7213402	25
26	142.0429320	.0070401	.0014889	.2114889	671.6330094	4.7283804	26
27	171.8719477	.0058183	.0012290	.2112290	813.6759414	4.7341986	27
28	207.9650567	.0048085	.0010147	.2110147	985.5478891	4.7390071	28
29	251.6377186	.0039740	.0008379	.2108379	1193.5129459	4.7429811	29
30	304.4816395	.0032843	.0006920	.2106920	1445.1506645	4.7462654	30
31	368.4227838	.0027143	.0005715	.2105715	1749.6323040	4.7489797	31
32	445.7915685	.0022432	.0004721	.2104721	2118.0550879	4.7512229	32
33	539.4077978	.0018539	.0003900	.2103900	2563.8466563	4.7530767	33
34	652.6834354	.0015321	.0003222	.2103222	3103.2544541	4.7546089	34
35	789.7469568	.0012662	.0002662	.2102662	3755.9378895	4.7558751	35
36	955.5938177	.0010465	.0002200	.2102200	4545.6848463	4.7569216	36
37	1156.2685195	.0008649	.0001818	.2101818	5501.2786640	4.7577864	37
38	1399.0849085	.0007148	.0001502	.2101502	6657.5471835	4.7585012	38
39	1692.8927393	.0005907	.0001241	.2101241	8056.6320920	4.7590919	39
40	2048.4002146	.0004882	.0001026	.2101026	9749.5248314	4.7595801	40

NOTE- **N IS EXPONENT N

22.00 PERCENT COMPOUND INTEREST FACTORS

	SINGLE PAYMENT		UNIFORM ANNUAL SERIES				
N PERIODS	COMPOUND AMOUNT FACTOR GIVEN P TO FIND S $(1 + I)^{**N}$	PRESENT WORTH FACTOR GIVEN S TO FIND P $\dfrac{1}{(1 + I)^{**N}}$	SINKING FUND FACTOR GIVEN S TO FIND R $\dfrac{I}{(1 + I)^{**N} - 1}$	CAPITAL RECOVERY FACTOR GIVEN P TO FIND R $\dfrac{I(1 + I)^{**N}}{(1 + I)^{**N} - 1}$	COMPOUND AMOUNT FACTOR GIVEN R TO FIND S $\dfrac{(1 + I)^{**N} - 1}{I}$	PRESENT WORTH FACTOR GIVEN R TO FIND P $\dfrac{(1 + I)^{**N} - 1}{I(1 + I)^{**N}}$	N PERIODS
1	1.2200000	.8196721	1.0000000	1.2200000	1.0000000	.8196721	1
2	1.4884000	.6718624	.4504505	.6704505	2.2200000	1.4915345	2
3	1.8158480	.5507069	.2696581	.4896581	3.7084000	2.0422414	3
4	2.2153346	.4513991	.1810201	.4010201	5.5242480	2.4936405	4
5	2.7027082	.3699993	.1292059	.3492059	7.7395826	2.8636398	5
6	3.2973040	.3032781	.0957644	.3157644	10.4422907	3.1669178	6
7	4.0227108	.2485886	.0727824	.2927824	13.7395947	3.4155064	7
8	4.9077072	.2037611	.0562990	.2762990	17.7623055	3.6192676	8
9	5.9874028	.1670173	.0441111	.2641111	22.6700127	3.7862849	9
10	7.3046314	.1368994	.0348950	.2548950	28.6574155	3.9231843	10
11	8.9116503	.1122127	.0278071	.2478071	35.9620469	4.0353970	11
12	10.8722134	.0919776	.0222848	.2422848	44.8736973	4.1273746	12
13	13.2641003	.0753915	.0179385	.2379385	55.7459107	4.2027661	13
14	16.1822024	.0617963	.0144907	.2344907	69.0100110	4.2645623	14
15	19.7422870	.0506527	.0117382	.2317382	85.1922134	4.3152150	15

n							n
16	4.3567336	104.9345004	.2295298	.0095298	.0415186	24.0855901	16
17	4.3907653	129.0200905	.2277507	.0077507	.0340316	29.3844199	17
18	4.4186601	158.4045104	.2263130	.0063130	.0278948	35.8489923	18
19	4.4415246	194.2535027	.2251479	.0051479	.0228646	43.7357706	19
20	4.4602661	237.9892733	.2242019	.0042019	.0187415	53.3576401	20
21	4.4756279	291.3469134	.2234323	.0034323	.0153619	65.0963209	21
22	4.4882196	356.4432343	.2228055	.0028055	.0125917	79.4175115	22
23	4.4985407	435.8607459	.2222943	.0022943	.0103211	96.8893641	23
24	4.5070006	532.7501099	.2218771	.0018771	.0084599	118.2050242	24
25	4.5139349	650.9551341	.2215362	.0015362	.0069343	144.2101295	25
26	4.5196188	795.1652636	.2212576	.0012576	.0056839	175.9363580	26
27	4.5242777	971.1016216	.2210298	.0010298	.0046589	214.6423568	27
28	4.5280965	1185.7439784	.2208434	.0008434	.0038188	261.8636752	28
29	4.5312266	1447.6076536	.2206908	.0006908	.0031301	319.4736838	29
30	4.5337923	1767.0813374	.2205659	.0005659	.0025657	389.7578942	30
31	4.5358953	2156.8392317	.2204636	.0004636	.0021030	475.5046310	31
32	4.5376191	2632.3438627	.2203799	.0003799	.0017238	580.1156498	32
33	4.5390321	3212.4595124	.2203113	.0003113	.0014129	707.7410927	33
34	4.5401902	3920.2006052	.2202551	.0002551	.0011582	863.4441331	34
35	4.5411395	4783.6447383	.2202090	.0002090	.0009493	1053.4018424	35
36	4.5419176	5837.0465808	.2201713	.0001713	.0007781	1285.1502478	36
37	4.5425554	7122.1968285	.2201404	.0001404	.0006378	1567.8833023	37
38	4.5430782	8690.C801308	.2201151	.0001151	.0005228	1912.8176288	38
39	4.5435067	10602.8977596	.2200943	.0000943	.0004285	2333.6375071	39
40	4.5438580	12936.5352667	.2200773	.0000773	.0003512	2847.0377587	40

NOTE- **N IS EXPONENT N

315

23.00 PERCENT COMPOUND INTEREST FACTORS

	SINGLE PAYMENT		UNIFORM ANNUAL SERIES				
N PERIODS	COMPOUND AMOUNT FACTOR GIVEN P TO FIND S $(1 + I)^{*N}$	PRESENT WORTH FACTOR GIVEN S TO FIND P $\dfrac{1}{(1 + I)^{*N}}$	SINKING FUND FACTOR GIVEN S TO FIND R $\dfrac{I}{(1 + I)^{*N} - 1}$	CAPITAL RECOVERY FACTOR GIVEN P TO FIND R $\dfrac{I(1 + I)^{*N}}{(1 + I)^{*N} - 1}$	COMPOUND AMOUNT FACTOR GIVEN R TO FIND S $\dfrac{(1 + I)^{*N} - 1}{I}$	PRESENT WORTH FACTOR GIVEN R TO FIND P $\dfrac{(1 + I)^{*N} - 1}{I(1 + I)^{*N}}$	N PERIODS
1	1.2300000	.8130081	1.0000000	1.2300000	1.0000000	.8130081	1
2	1.5129000	.6609822	.4484305	.6784305	2.2300000	1.4739903	2
3	1.8608670	.5373839	.2671725	.4971725	3.7429000	2.0113743	3
4	2.2888664	.4368975	.1784514	.4084514	5.6037670	2.4482718	4
5	2.8153057	.3552012	.1267004	.3567004	7.8926334	2.8034730	5
6	3.4628260	.2887815	.0933887	.3233887	10.7079391	3.0922545	6
7	4.2592760	.2347817	.0705678	.3005678	14.1707651	3.3270361	7
8	5.2389094	.1908794	.0542592	.2842592	18.4300411	3.5179156	8
9	6.4438586	.1551865	.0422494	.2722494	23.6689505	3.6731021	9
10	7.9259461	.1261679	.0332085	.2632085	30.1128091	3.7992700	10
11	9.7489137	.1025755	.0262890	.2562890	38.0387552	3.9018455	11
12	11.9911638	.0833947	.0209259	.2509259	47.7876689	3.9852403	12
13	14.7491315	.0678006	.0167283	.2467283	59.7788328	4.0530409	13
14	18.1414318	.0551224	.0134178	.2434178	74.5279643	4.1081633	14
15	22.3139611	.0448150	.0107910	.2407910	92.6693961	4.1529783	15

316

N			
16	4.1894132	114.9833572	.2386969
17	4.2190352	142.4295293	.2370210
18	4.2431180	176.1883211	.2356757
19	4.2626976	217.7116349	.2345932
20	4.2786159	268.7853109	.2337204
21	4.2915577	331.6059325	.2330156
22	4.3020794	408.8752969	.2324457
23	4.3106337	503.9166152	.2319845
24	4.3175883	620.8174367	.2316108
25	4.3232425	764.6054472	.2313079
26	4.3278395	941.4647000	.2310622
27	4.3315768	1159.0015810	.2308628
28	4.3346153	1426.5719447	.2307010
29	4.3370856	1755.6834919	.2305696
30	4.3390940	2160.4906951	.2304629
31	4.3407268	2658.4035549	.2303762
32	4.3420543	3270.8363726	.2303057
33	4.3431336	4024.1287383	.2302485
34	4.3440111	4950.6783481	.2302020
35	4.3447244	6090.3436810	.2301642
36	4.3453044	7492.1112728	.2301335
37	4.3457759	9216.2968656	.2301085
38	4.3461593	11337.0451446	.2300882
39	4.3464710	13945.5655279	.2300717
40	4.3467244	17154.0455993	.2300583

N			
16	27.4461722	.0364350	.0086969
17	33.7587917	.0296219	.0070210
18	41.5233138	.0240829	.0056757
19	51.0736760	.0195796	.0045932
20	62.8206215	.0159183	.0037204
21	77.2693645	.0129417	.0030156
22	95.0413183	.0105217	.0024457
23	116.9008215	.0085543	.0019845
24	143.7880104	.0069547	.0016108
25	176.8592528	.0056542	.0013079
26	217.5368810	.0045969	.0010622
27	267.5703636	.0037373	.0008628
28	329.1115473	.0030385	.0007010
29	404.8072031	.0024703	.0005696
30	497.9128599	.0020084	.0004629
31	612.4328176	.0016328	.0003762
32	753.2923657	.0013275	.0003057
33	926.5496098	.0010793	.0002485
34	1139.6560201	.0008775	.0002020
35	1401.7769047	.0007134	.0001642
36	1724.1855927	.0005800	.0001335
37	2120.7482791	.0004715	.0001085
38	2608.5203833	.0003834	.0000882
39	3208.4800714	.0003117	.0000717
40	3946.4304878	.0002534	.0000583

NOTE- **N IS EXPONENT N

24.00 PERCENT COMPOUND INTEREST FACTORS

	SINGLE PAYMENT		UNIFORM ANNUAL SERIES				
N PERIODS	COMPOUND AMOUNT FACTOR GIVEN P TO FIND S $(1+I)^{**N}$	PRESENT WORTH FACTOR GIVEN S TO FIND P $\dfrac{1}{(1+I)^{**N}}$	SINKING FUND FACTOR GIVEN S TO FIND R $\dfrac{I}{(1+I)^{**N}-1}$	CAPITAL RECOVERY FACTOR GIVEN P TO FIND R $\dfrac{I(1+I)^{**N}}{(1+I)^{**N}-1}$	COMPOUND AMOUNT FACTOR GIVEN R TO FIND S $\dfrac{(1+I)^{**N}-1}{I}$	PRESENT WORTH FACTOR GIVEN R TO FIND P $\dfrac{(1+I)^{**N}-1}{I(1+I)^{**N}}$	N PERIODS
1	1.2400000	.8064516	1.0000000	1.2400000	1.0000000	.8064516	1
2	1.5376000	.6503642	.4464286	.6864286	2.2400000	1.4568158	2
3	1.9066240	.5244873	.2647183	.5047183	3.7776000	1.9813031	3
4	2.3642138	.4229736	.1759255	.4159255	5.6842240	2.4042767	4
5	2.9316251	.3411077	.1242477	.3642477	8.0484378	2.7453844	5
6	3.6352151	.2750869	.0910742	.3310742	10.9800628	3.0204713	6
7	4.5076667	.2218443	.0684216	.3084216	14.6152779	3.2423156	7
8	5.5895067	.1789067	.0522932	.2922932	19.1229446	3.4212222	8
9	6.9309883	.1442796	.0404654	.2804654	24.7124513	3.5655018	9
10	8.5944255	.1163545	.0316021	.2716021	31.6434396	3.6818563	10
11	10.6570876	.0938343	.0248522	.2648522	40.2378651	3.7756906	11
12	13.2147887	.0756728	.0196483	.2596483	50.8949527	3.8513634	12
13	16.3863379	.0610264	.0155983	.2555983	64.1097414	3.9123898	13
14	20.3190590	.0492149	.0124230	.2524230	80.4960793	3.9616047	14
15	25.1956332	.0396894	.0099191	.2499191	100.8151384	4.0012941	15

N						N
16	31.2425852	.0320076	.0079358	.2479356	126.0107716	4.0333017	16
17	38.7408056	.0258126	.0063592	.2463592	157.2533568	4.0591143	17
18	48.0385990	.0208166	.0051022	.2451022	195.9941624	4.0799309	18
19	59.5678627	.0167876	.0040978	.2440978	244.0327614	4.0967184	19
20	73.8641498	.0135384	.0032936	.2432938	303.6006241	4.1102568	20
21	91.5915457	.0109180	.0026493	.2426493	377.4647739	4.1211748	21
22	113.5735167	.0088049	.0021319	.2421319	469.0563196	4.1299707	22
23	140.8311607	.0071007	.0017164	.2417164	582.6298363	4.1370804	23
24	174.6306393	.0057264	.0013822	.2413822	723.4609971	4.1428068	24
25	216.5419927	.0046180	.0011135	.2411135	898.0916364	4.1474248	25
26	268.5120710	.0037242	.0008972	.2408972	1114.6336291	4.1511491	26
27	332.9549680	.0030034	.0007230	.2407230	1383.1457001	4.1541525	27
28	412.8641603	.0024221	.0005827	.2405827	1716.1066681	4.1565746	28
29	511.9515588	.0019533	.0004697	.2404697	2128.9648284	4.1585279	29
30	634.8199329	.0015752	.0003787	.2403787	2640.9163873	4.1601031	30
31	787.1767168	.0012704	.0003053	.2403053	3275.7363202	4.1613735	31
32	976.0991289	.0010245	.0002461	.2402461	4062.9130370	4.1623980	32
33	1210.3629198	.0008262	.0001985	.2401985	5039.0121659	4.1632242	33
34	1500.8500206	.0006663	.0001600	.2401600	6249.3750658	4.1638905	34
35	1861.0540255	.0005373	.0001290	.2401290	7750.2251063	4.1644278	35
36	2307.7069916	.0004333	.0001040	.2401040	9611.2791319	4.1648611	36
37	2861.5566696	.0003495	.0000839	.2400839	11918.9861235	4.1652106	37
38	3548.3302704	.0002818	.0000677	.2400677	14780.5427932	4.1654924	38
39	4399.9295352	.0002273	.0000546	.2400546	18328.8730635	4.1657197	39
40	5455.9126237	.0001833	.0000440	.2400440	22728.8025988	4.1659030	40

NOTE- **N IS EXPONENT N

25.00 PERCENT COMPOUND INTEREST FACTORS

	------SINGLE PAYMENT------		------------------------------UNIFORM ANNUAL SERIES------------------------------				
N PERIODS	COMPOUND AMOUNT FACTOR GIVEN P TO FIND S $(1 + I)^{**N}$	PRESENT WORTH FACTOR GIVEN S TO FIND P $\dfrac{1}{(1 + I)^{**N}}$	SINKING FUND FACTOR GIVEN S TO FIND R $\dfrac{I}{(1 + I)^{**N} - 1}$	CAPITAL RECOVERY FACTOR GIVEN P TO FIND R $\dfrac{I(1 + I)^{**N}}{(1 + I)^{**N} - 1}$	COMPOUND AMOUNT FACTOR GIVEN R TO FIND S $\dfrac{(1 + I)^{**N} - 1}{I}$	PRESENT WORTH FACTOR GIVEN R TO FIND P $\dfrac{(1 + I)^{**N} - 1}{I(1 + I)^{**N}}$	N PERIODS
1	1.2500000	.8000000	1.0000000	1.2500000	1.0000000	.8000000	1
2	1.5625000	.6400000	.4444444	.6944444	2.2500000	1.4400000	2
3	1.9531250	.5120000	.2622951	.5122951	3.8125000	1.9520000	3
4	2.4414063	.4096000	.1734417	.4234417	5.7656250	2.3616000	4
5	3.0517578	.3276800	.1218467	.3718467	8.2070313	2.6892800	5
6	3.8146973	.2621440	.0888195	.3388195	11.2587891	2.9514240	6
7	4.7683716	.2097152	.0663417	.3163417	15.0734863	3.1611392	7
8	5.9604645	.1677722	.0503985	.3003985	19.8418579	3.3289114	8
9	7.4505806	.1342177	.0387562	.2887562	25.8023224	3.4631291	9
10	9.3132257	.1073742	.0300726	.2800726	33.2529030	3.5705033	10
11	11.6415322	.0858993	.0234929	.2734929	42.5661287	3.6564026	11
12	14.5519152	.0687195	.0184476	.2684476	54.2076609	3.7251221	12
13	18.1898940	.0549756	.0145434	.2645434	68.7595761	3.7800977	13
14	22.7373675	.0439805	.0115009	.2615009	86.9494702	3.8240781	14
15	28.4217094	.0351844	.0091169	.2591169	109.6868377	3.8592625	15

25.00%

N							N
16	3.8874100	138.1085472	.2572407	.0072407	.0281475	35.5271368	16
17	3.9099280	173.6356839	.2557592	.0057592	.0225180	44.4089210	17
18	3.9279424	218.0446049	.2545862	.0045862	.0180144	55.5111512	18
19	3.9423539	273.5557562	.2536556	.0036556	.0144115	69.3889390	19
20	3.9538831	342.9446952	.2529159	.0029159	.0115292	86.7361738	20
21	3.9631065	429.6808690	.2523273	.0023273	.0092234	108.4202172	21
22	3.9704852	538.1010862	.2518584	.0018584	.0073787	135.5252716	22
23	3.9763882	673.6263578	.2514845	.0014845	.0059030	169.4065895	23
24	3.9811105	843.0329473	.2511862	.0011862	.0047224	211.7582368	24
25	3.9848884	1054.7911841	.2509481	.0009481	.0037779	264.6977960	25
26	3.9879107	1319.4889801	.2507579	.0007579	.0030223	330.8722450	26
27	3.9903286	1650.3612251	.2506059	.0006059	.0024179	413.5903063	27
28	3.9922629	2063.9515314	.2504845	.0004845	.0019343	516.9878828	28
29	3.9938103	2580.9394142	.2503875	.0003875	.0015474	646.2348536	29
30	3.9950482	3227.1742678	.2503099	.0003099	.0012379	807.7935669	30
31	3.9960386	4034.9678347	.2502478	.0002478	.0009904	1009.7419587	31
32	3.9968309	5044.7097934	.2501982	.0001982	.0007923	1262.1774484	32
33	3.9974647	6306.8872418	.2501586	.0001586	.0006338	1577.7218104	33
34	3.9979718	7884.6090522	.2501268	.0001268	.0005071	1972.1522631	34
35	3.9983774	9856.7613153	.2501015	.0001015	.0004056	2465.1903288	35
36	3.9987019	12321.9516441	.2500812	.0000812	.0003245	3081.4879110	36
37	3.9989615	15403.4395551	.2500649	.0000649	.0002596	3851.8598888	37
38	3.9991692	19255.2994439	.2500519	.0000519	.0002077	4814.8248610	38
39	3.9993354	24070.1243048	.2500415	.0000415	.0001662	6018.5310762	39
40	3.9994683	30088.6553811	.2500332	.0000332	.0001329	7523.1638453	40

NOTE- **N IS EXPONENT N

BIBLIOGRAPHY

COMPUTER APPLICATIONS

Cobb, D. F. and G. B. Cobb. *SuperCalc SuperModels for Business*. Que Corp., Indianapolis, IN, 1983.

Nevison, J. N. *Executive Computing*. Addison-Wesley, Reading, MA, 1981.

Spans, L. E. *Investment Analysis by Microcomputer*. TAB Books, Blue Ridge Summit, PA, 1982.

Sternberg, C. D. *BASIC Computer Programs for Business* (Vols. 1 and 2). Hayden, Rochelle Park, NJ. Vol. 1, 1980; Vol. 2, 1982.

Trost, S. R. *Doing Business with SuperCalc*. Sybex, Berkeley, CA, 1983.

Watson, W. S. *55 Advanced Computer Programs in BASIC*. TAB Books, Blue Ridge Summit, PA, 1981.

COST AND PROFIT ANALYSIS

Milling, B. E. *Cash Flow Problem Solver*. Chilton, Radnor, PA, 1981.

Park, W. R. *Construction Bidding for Profit*. Wiley, New York, 1979.

Pritchard, R. E. and B. M. Bradway. *Strategic Planning and Control Techniques for Profit*. Prentice-Hall, Englewood Cliffs, NJ, 1981.

Tucker, S. A. *Profit Planning Decisions with the Break-Even System*. Thomond Press, New York, 1980.

DEPRECIATION

Economic Recovery Tax Act of 1981: An Analysis of the New Legislation. Arthur Young & Co., 1981.

1981–1983 Federal Tax Highlights. Arthur Young & Co., 1982.

ENGINEERING ECONOMICS

Blank, L. T. and A. J. Tarquin. *Engineering Economy*, 2nd ed. McGraw-Hill, New York, 1983.

Jelen, F. C. (Ed.). *Cost and Optimization Engineering*. McGraw-Hill, New York, 1970.

Jones, B. W. *Inflation in Engineering Economics*. Wiley-Interscience, New York, 1982.

GENERAL INTEREST

Altman, Edward I. (Ed.). *Financial Handbook*, 5th ed. Wiley, New York, 1981.

Daniells, L. M. *Business Information Sources*. University of California Press, Berkeley, CA, 1976.

Klug, J. R. *The Basic Book of Business*. CBI, Boston, 1977.

Kroeber, D. W. and R. L. LaForge. *Manager's Guide to Statistics and Quantitative Methods*. McGraw-Hill, New York, 1980.

Park, W. R. and S. C. Park. *How to Succeed in Your Own Business*. Wiley, New York, 1978.

Symonds, C. W. *Basic Financial Management*. AMACOM, New York, 1978.

INVESTMENT EVALUATION AND FINANCIAL ANALYSIS

Cohen, J. B. and S. M. Robbins. *The Financial Manager*. Harper & Row, New York, 1966.

Park, W. R. and J. B. Maillie. *Strategic Analysis for Venture Evaluation*. Van Nostrand-Reinhold, New York, 1982.

Shuckett, D. H. and E. J. Mock. *Decision Strategies in Financial Management.* AMACOM, New York, 1973.

Schwartzman, S. D. and R. E. Ball. *Elements of Financial Analysis.* Van Nostrand-Reinhold, New York, 1977.

Spiro, H. T. *Finance for the Nonfinancial Manager.* Wiley, New York, 1977.

Sweeny, Allen. *Accounting Fundamentals for Nonfinancial Executives and Managers.* McGraw-Hill, New York, 1977.

Index